Inventoried $51.99

S0-ERD-555

Paola Free Library
101 E. Peoria
Paola, KS 66071

PAOLA FREE LIBRARY
101 E PEORIA
PAOLA KANSAS 66071

2001   GAYLORD

Female Strategies

EVELYN SHAW
AND
JOAN DARLING

A TOUCHSTONE BOOK
Published by Simon & Schuster, Inc.
NEW YORK

Copyright © 1985 by Evelyn Shaw and Joan Darling

All rights reserved
including the right of reproduction
in whole or in part in any form

First Touchstone Edition, 1986

Published by Simon & Schuster, Inc.
Simon & Schuster Building
Rockefeller Center
1230 Avenue of the Americas
New York, New York 10020

Published by arrangement with Walker and Company

TOUCHSTONE and colophon are registered trademarks
of Simon & Schuster, Inc.

Manufactured in the United States of America

10 9 8 7 6 5 4 3 2 1 Pbk.

Library of Congress Cataloging in Publication Data

ISBN 0-671-42354-1 Pbk.

We are grateful to the following companies for their permission to use material quoted in this book:

W. H. Freeman and Company for Daniel Lehrman's article, "The Reproductive Behavior of Ring Doves," in the November 1964 issue of *Scientific American*

Warner Bros. Inc. for Cole Porter's "Let's Do It" © 1928 (Renewed) Warner Bros. Inc. All Rights Reserved. Used by Permission.

Crazy Crow Music for Joni Mitchell's "Barangrill" © 1972 Crazy Crow Music. All Rights Reserved. Used by Permission.

*To our children, grandchildren,
and husbands*

CONTENTS

INTRODUCTION: *Shattering Stereotypes*   1

**1. SPECIES SURVIVAL:
THE PROMISCUOUS SHINER PERCH**   5
A Biological Nonconformist? · *Promiscuity: The Best Insurance Policy* · *The Persistence of a Sexist Biology* · *Liberated Ladies of the Animal Kingdom*

**2. WHERE FEMALES END
AND MALES BEGIN**   19
Sex: Playing to Win · *A Sperm Is a Sperm, But an Egg Is—?* · *Gender: Sex Is All in Your Head* · *The Queen Bee and Her Genetic Monarchy* · *Having It Both Ways*

3. TRANSITION   47
Young Love and Perfect Timing · *The Menstrual Cycle: Harmony of Body and Brain* · *Twenty-eight Days—a Close-up*

4. TIMING IS EVERYTHING   59
The Ring Dove: Feedback Systems at Work · *Eggs Welcome Here* · *The Red-Deer Harem* · *Eggs On Demand* · *Being Female Without Reproducing*

5. IN THE HEAT OF ESTRUS   75
Speculations About Our Loss of Estrus · *Estrus: Biological Excess Baggage?*

6. **A MILLION WAYS
   TO CATCH A MALE** 91
   *The Lek: A Sexual Supermarket · When She
   Does the Asking · Fathers Who Get Custody
   · Bargaining For Sex · Long-Term Pair Bonding*

7. **THE TYRANT EMBRYO** 121
   *Guarding the Eggs · Toting the Eggs · Parasitic
   Embryos · Giving Birth: A Solitary Experience*

8. **THE MYTH OF MATERNALISM** 133
   *Binding Mother and Infant · From Coatis to
   Samoan Natives — Sharing Maternity · Excluding
   the Father*

9. **NEW ROLES, NEW CHOICES** 151

   CHAPTER NOTES 155

   INDEX 159

FEMALE STRATEGIES

INTRODUCTION

Shattering Stereotypes

TECHNOLOGY HAS FREED women to enter the habitats of many other species—to soar on thermal updrafts, like an eagle; to dive into the depths of the sea, like a gray whale; to tread cautiously among the rocky outcroppings of a mountain, like a bighorn sheep; and to paddle lazily on the ocean, like a sea otter. But unlike us, eagles don't strap on scuba gear to dive into the ocean, gray whales don't climb Mount Everest, and sea otters don't pilot airplanes. These species are confined to their respective habitats, whereas ours is the only one, among millions, where a woman can drink hot coffee in an igloo in Alaska in the morning and sip iced tea on a lanai in Hawaii that afternoon, without losing her species identity.

Yet as we travel through the spaces occupied by the many diverse species wrought by evolution, we are aware of their females only when they affect our lives directly—when bitten by a mosquito, for instance, or confronted with the threatening grimace of a cornered bitch. We are unaware that we are connected to the females of all species by a common denominator: We are all producers of eggs, the raw material of species survival. We are unaware that the mosquito, by biting, is carrying out her reproductive mandate by gathering protein to make her eggs and that the bitch, by snarling, is carrying out her mandate to protect her young. In all but asexual species, the burden to perpetuate life on earth falls to the female; appearances, behaviors, and physiologies that are unique to a species evolve in response to the female's reproductive demands, in response to her need to get her eggs fertilized and to maximize the chances of her embryo surviving to reproductive age. Species' life-styles revolve around females and their reproductive mandate.

In our continuing search for comprehension of human behavior, we often turn to the behavior of other animals to explain ourselves to ourselves, to justify our acts, or to give evolutionary rationale to altruistic as well as destructive tendencies. We willingly embrace the postulate that behavior results from biologically determined directives. We happily analogize the territorial tendencies of a coral-reef fish to our own territorial tendencies, and when we discover that estrogen levels control the behavior of a female mammal in heat, we assume that our courting behavior is biologically instigated as well.

We strive to derive knowledge of ourselves from animal models, thinking that we are doing so with an unbiased eye, but we delude ourselves. For as we examine their behaviors we tend to project our own stereotypical expectations of female and male behavior on them. Essentially, by burdening these models with our biases, preconceptions, and expectations, we distort their reality and we distort what they may have to tell us.

When we study, observe, analyze, and experiment on animals, we soon discover that laboring under stereotypical expectations hugely constrains the kinds of questions we want to ask, to find out about ourselves. If we push aside preconceptions, open ourselves up and let the females of other species "talk" to us, we soon become aware of how superficial and trite our stereotypes are. For example, within the stereotypical mode, traits classified as masculine aptly describe the females of some species who are brawny, tower above their mates, sport brilliant colors, have huge sexual appetites, or even hang a penis and scrotum on their abdomens. And, conversely, words traditionally used to describe feminine qualities make a mockery of the actual behavior of some females. Can you envision a female black widow spider *coyly* raising her abdomen to invite copulation or *gently* embracing the male as she proceeds to eat him? Can you imagine a *sultry* female gorilla lowering her eyelids and *subtly shrinking away* as the male advances toward her, intent on copulating? Can you believe that a mother grizzly bear stands *passively* by as a predator consumes her cubs?

Oblivious to our preconceptions and expectations females of many species have the evolutionary freedom to acquire any appearance or employ any behavior to enhance their reproductive success. Females are many-faceted, fashioning themselves into a variety of guises and styles, ranging from the switch-hitting female water flea to the sexually parasitic female salamander; from the female hare, who

INTRODUCTION: SHATTERING STEREOTYPES

readies her eggs at any sexual tryst, to the female human, who has many trysts but never knows when her egg is ready.

These fascinating and astonishingly diverse styles are a result of biological evolution, which keeps a strict concordance between different reproductive needs and life-styles. Even among humans, basic biology has been modified to suit a culturally shaped life-style, although most women are unaware of these profound evolutionary specializations. Society's expectations have wrapped the human female—like an Egyptian mummy—in so many layers of culturally fabricated taboos, rites, distortions, and superstitions that, as we shall see, she is oblivious to the freedom lying dormant within her, endowed by her biology.

Even among odd and bizarre species, the way a female accomplishes her reproductive mandate is irrelevant as long as it is fulfilled. Indeed, these species reaffirm our premise that evolutionary options are almost limitless.

The emphasis of scientific research until about ten years ago was on the male, and the major research thrust pursued his behaviors, treating the female only as an incidental sperm receptacle. Now, the analysis of female biology, including physiology and behavioral traits, is the focus of scientific work in laboratories around the world, rectifying the decades during which females were ignored by scientists. The outcome of recent research is the shattering of stereotypes, like the discovery of the aggressive role that the female plays in courtship and mating, and her role as the pivotal member of animal societies. For example, in baboon tribes, the dominant male was always assumed to be the dominant member of the group, but careful observations recently revealed that the dominant female, not the male, is the leader—baboon society is a matriarchy, not the patriarchy it was assumed to be. Furthermore, among other species to be discussed in this book—such as the horse or monkey—it is the female who initiates courtship, using a series of subtle cues, such as a swish of the tail or a heady perfume, cues so subtle that they were overlooked by earlier researchers but clearly not by the male of the species. In addition, our own research has revealed the complexities of the female strategies of courtship and mating in so arcane a species as the shiner perch—which is promiscuous, not content with merely one mate. She solicits the attentions of many, a behavioral strategy suiting her system of delayed fertilization. Other recent and significant research studies strike at the very roots of biological assumptions. At one time the relationship between a female's behavior and her hormone level was thought to be a one-way street, with "raging" hormones dictating behavior. This relationship now proves to be a two-way street: hormone levels not only create behaviors, but behaviors alter hormone levels.

The shelves and shelves of books about women in culture, women in myth, women in primitive societies — on women's anthropology, psychology, and sociology — all tend to shy away from the woman as a biological entity. In contrast, in some books, culture is removed entirely as an influence and the pendulum swings completely in the opposite direction. The female human is made into a pawn of her genes, a slave to her hormones, a creature in whom behaviors are etched in the neural network of her brain. Either view is extreme. Each distorts the interdependence of the environment, the brain and the rest of the body.

We need to appreciate the evolutionary basis of femininity, to compare our biological selves to females of other species. From such comparisons women will begin to recognize where their own biology ends and cultural influences begin. They will start to think about how to use the scientific knowledge, gleaned from animal studies, to liberate themselves from the crushing weight of cultural dictates. With a new understanding of the biological basis of femininity, can new cultural forms of femininity be far behind?

1

SPECIES SURVIVAL: THE PROMISCUOUS SHINER PERCH

IN THE 1950S AND '60S, one of us, as a scientist in the department of animal behavior at the American Museum of Natural History, believed as did her colleagues that sexual behavior in virtually all animals was the male mandate. The peremptory male was the initiator; he sought out the female, courted and copulated at the time of his choosing, and then, having satisfied his sexual desires (at least with that female), he crept, swam, or flew away as species limbs dictated. We could not resist the male, a wonderful creature to watch. Indeed, his appearance and behavior are specifically designed to get attention from potential mates and competing suitors. For example, a jewel fish male, readying himself for his tryst, changes his grayish body into a stunning ruby red and flashes his gemlike sides to entice any passing females into his staked-out gravel heap. An amorous tomcat howls out his interest in a sexual encounter and improves his chances of copulation by actively prowling in search of the body perfumes given off by a female in heat. The white-crowned sparrow sings his melodious and ingratiating song from the treetops, attesting to his superior attributes as a potential mate and advertising his leafy penthouse with room for a nest. Luckily for him, his song lures a female. The Norway rat skips the advertising and scrambles through urban sewers in hot pursuit of the telltale odors of a female in estrus. However, despite the singlemindedness of his search, he is keenly alert and ready to do battle with another male, who may also be in pursuit of his potential prize.

As scientists, we are drawn to the male's activities; he is always doing something, providing the substance for analysis and quantification and the matrix for theoretical exposition of behavioral form and function. He provides us with data, the nut of scientific research. His behavior can be described, counted, timed, arranged in sequences, and because it is predictable and consistent, it can be manipulated. Through manipulation, we can elucidate cause and effect. In our research programs, we danced to the music of nature/nurture controversy using the steps of mating to answer our questions about the extent to which gene power and environmental power influenced an animal's behavior.

We studied male cats, watching them grow from infancy through prepubescence and finally into sexual maturity, constantly recording changes in behavior. Using their normal behavior as the standard, we experimentally induced them to answer questions about the biological and/or environmental bases of sexual behavior. We investigated the impact of social environment, hormones and the nervous system on the male's sexual arousal and copulation. We controlled the amount of contact he had with others of his own species, we regulated his hormone level, and we extirpated parts of his brain as well as stimulating certain regions with chemicals and electric shocks. When such experimentally manipulated males were ready to be tested, we introduced the female. In experiments we watched the pair closely, but we recorded only the male's behavior, counting how often he looked at her, pursued, approached, nipped, mounted, attempted to copulate. We clocked the minutes, even the seconds of each act and we chronicled the sequences. Then, when the experiments were deemed complete, we compared the sexual behavior of manipulated males to the behavior of normal, untouched males, arriving at profound conclusions. That male cats who were castrated (the source of testosterone removed) before sexual maturity were basically uninterested in females, but that sexually experienced, castrated cats continued their pursuit of the female, albeit less ardently; that removal of certain parts of the brain frequently left the sexuality of the animal intact; and that being raised in isolation without the contact of other cats led to socially unacceptable, disjointed, bizarre behavior—so bizarre that it was impossible to know what their normal sexual appetites might have been.

We ignored the female. We made her a feline robot. By removing her ovaries, we neutered her own sexuality, and then artificially created her sexual heat with megadoses of estrogen injected into her flanks. We directed her to be receptive under the baton of estrogen. By using the same scientifically created sexual robots over and over, researchers didn't have to deal with complications generated by the idiosyncracies of intact females who might or might not be in heat on the day of testing. They had one less variable to contend with when quantitatively analyzing the sexual behavior of the male.

All over the United States and Europe, sex researchers were on the same bandwagon, rolling to the beat of the male drum, each one reaffirming the others' experimental results, published in a vast literature, iterating and reiterating that sexual behavior was the male imperative. Because of the focus of the research, and because the female's inclinations were minimized through manipulation, neutering, and so forth, the conclusions were universally clear: The male was stereotyped as the active partner, the instigator, the broadcaster, whereas the female was stereotyped as passive, a receiver without a mind of her own.

This emphasis on the male, reinforcing our gender stereotypes, led to an unfortunate imbalance, a diminishing of the female and an exaggeration of the male contribution to mating. The experiments were like watching *Romeo and Juliet* through a telescope fixed on Romeo, ignoring everything that Juliet says or does. How can the story of courtship be complete if only half the dialogue has been heard? Only in recent years have scientists discovered that the female, free from tranquilizers, ovariectomies, and hormone injections, has the major lines to say. Research of the 1970s and early '80s reveals her to be far from passive, and far from being a mute, constantly accessible sexual partner.

A Biological Nonconformist?

In our studies done at Stanford University and the Bodega Marine Laboratory during the 1970s, a small silvery fish became the spokeswoman for the generations of females overlooked and ignored in the mating game. Although a fish, and limited in her expressive abilities, she nonetheless pioneered a scientific frontier. Commonly called the shiner perch, and known in the scientific community as *Cymatogaster aggregata*, she inhabits shallow waters found along the entire Pacific coast of North America from the fjords of Alaska to the beaches of Baja.

Most fish reproduce by spawning eggs into a watery milieu, but this fish gives birth. Only four inches long, she gives birth during the spring and summer to about ten large young, who emerge from their mother's ovary well equipped to make their own way in the world. Soon after they are born, females in this species mate, but their eggs are not immediately fertilized. Instead, the female stashes the male's sperm within her ovaries. There she feeds the sperm nutritious tidbits to keep them lively and virile for several months, a most unusual procedure. But this is the least of her unusual behaviors.

During the breeding season of June, July, and August, the shiner perch female is in reproductive limbo. Her gonads, collapsed after the trauma of giving birth, are in no condition to make eggs, and they will not rev up again until winter. In contrast, the male is eager to mate. He dons lover's colors, by darkening his silvery skin into a dusky gray. He establishes and patrols a territory, making amorous advances to any passing silvery fish. Yet he and the females are out of sync. The sexes seem to have set their biological alarm clocks to different time zones — warm waters and longer days turn on the male, while short days and cool water turn on the female. The female's ovaries are shut down during the male's moments of intense passion.

She has neither eggs nor any stimulating ovarian hormones; so according to the biological orthodoxy that hormones create female receptivity, she should not be aroused by, or even vaguely interested in the sexy, soliciting male. Yet she mates. In his sexual high, does the male rape her, or is she being a biological nonconformist?

To ferret out the clues about this unusual system where the sexes are physiologically out of sync while mating, we gathered into our nets hundreds of shiner perch, prodding them to reveal their sexual secrets. From the males, milky white sperm flowed, telling us that they were sexually ready. Some females were still pregnant; some had given birth, but their empty ovaries indicated no matings; other postpartum females had already stashed lively sperm in their ovarian banks.

We subjected these females to scientific scrutiny, pairing them with males and recording their courtship. Since part of the challenge and fun of science is prediction of results, we hypothesized the responses of the female. To begin with, we couldn't be sure that any females would be interested in courtship at this time, because they all lacked eggs. We guessed that if any were sexually active they should be the postpartum unmated females; we predicted that the pregnant females and the already mated females might not court and perhaps would try to escape from the amorous gestures of the sexually ready, dusky-colored males. After all, pregnant females, like all good mothers, should be concerned with giving birth. Furthermore, according to traditional biological dogma, mated females should be satisfied with the millions of sperm supplied by one male. Much to our surprise—the cliché of science—both kinds of postpartum females, those carrying sperm as well as those who were uninseminated, were very attentive to the male's courting overtures, his lateral displays, his gyrations as he swam in figure-eight patterns in front of them, even though these females were eggless. And not only were they receptive to the male's advances, but they even pursued him, initiating the passes, the chases, and the side-by-side swimming. Only the pregnant female, true to our expectation, tried to avoid the male and escape, scurrying up the walls of the confining aquarium tank in an attempt to get out.

The female that particularly intrigued us was the one who had already stashed sperm in her ovaries. According to biological dogma, that the sperm from one male is enough for any female, she didn't need any more sperm. Yet she actively pursued several males in sequence, making as many advances to them as they made to her, and mating with them in our tanks. Certainly she was genuinely interested in a male's advances, but she was flouting that biological stereotype dictating that one male is enough for any female. Was this because she had nothing better to do in a small aquarium? Confinement may force animals into abnormal behavior patterns. Clearly it is imperative to study the female in her natural habitat. Does the female mate with more than one male in the ocean, where life makes more serious demands on her time?

Promiscuity: The Best Insurance Policy

The water where the shiner perch lives is so murky that one's own hand is barely visible, so we could never observe females mating on their turf. We had to resort to indirect means by employing a technique that is used in human paternity suits. Human courts use blood types and other biological evidence to try to match fathers to their progeny, and we did likewise. We put the shiner perch on trial, in the piscine equivalent of a paternity suit.

Although we would not have been surprised to find that a few females mated more than once, as suggested by the laboratory work, we were amazed to discover that two or more fathers was the rule rather than the exception. Not only did these females continue to mate after it seemed unnecessary, but they were downright promiscuous.

Without the benefit of ovarian hormones to trigger their sex behavior, they mate with more than one male, even though one male can supply an excess of sperm to fertilize 40 or 50 eggs. They mate when the chance arises, and mating has to mesh with their peculiar life-style, in which females and males come together once a year in the same warm, shallow bays. Pregnant females migrate to the warmer waters to give birth, congregating in vast schools. After dropping their young in lush, protective vegetation, they swim back toward deeper waters, encountering on the way dark-colored males vigorously demonstrating their sexual readiness as they guard a small bit of barren territory. Taking advantage of the large number of males, the female mates promiscuously, gathering millions of sperm into her ovary, an insurance policy that guarantees the survival of some sperm when her eggs are ripened. Sperm cells, being fragile, do not easily survive their four months in storage, despite the nurturant environment of ovarian tissue.

After mating, the female swims out of the bay into the Pacific Ocean. Why does she bother with sperm storage and the responsibility of keeping these delicate gametes alive and well? Why not use the sperm right away instead of chancing their four-month survival? But another factor enters into her reproductive scheme: timing the birth of her young. If she started her eggs on their developmental road in August, their four-to-five-month gestation would end in December. Her winter home, far from thick, lush vegetation, is clearly not a suitable place to start life, for there is little to eat. Thus, by delaying egg maturation and fertilization, she gives birth to her young in the early summer, after she has migrated to the pleasant climes of a warm, vegetation-rich bay, thereby ensuring her progeny of plentiful food of the right size.

The shiner perch, lowly creature that she is, cannot read, does not know her parents, is not subjugated to the clout of culture, and has not discovered that promiscuity among females is a no-no. She follows her biological tendencies, doing what she must do to reproduce. Not caught up in what's expected of her just because she is female, she is, however, caught up in making her best adjustments to the demands of life-style and survival. If she has to pursue a male when she has no eggs, she will. If she improves the chances of becoming a mother by being promiscuous, so be it. Obviously they have not read the biological book decreeing that feminine sexual appetite waxes and wanes with the ovarian hormone estrogen and that females indulge in mating as little as possible. These females ignore stereotypes. They adapt their behavior to suit their reproductive needs and their life-style.

Even more recently, several researchers, looking at female strategies, have shown that promiscuity is much more common than expected. It is practiced by lots of females, in lots of species, running the gamut from butterflies to salamanders, from ground squirrels to swordtail fish, from mosquitoes to garter snakes, from fruit flies to chimpanzees. Despite the fact that it is usually considered to be a male prerogative by scientists and historians, the mating strategy of promiscuity is embraced by females to meet an environmental demand specific to their species life-style. This discovery runs headlong into our classic stereotype of the female as a passive, relatively inert partner in sex, the awaiter instead of the pursuer.

The Persistence of a Sexist Biology

Despite the recent discoveries of these "unusual" female behaviors, the feminine stereotype is so tightly etched in scientific thought that a female who is brightly colored, promiscuous, and a gadabout is described as showing "sex role reversal"—in other words, taking the expected masculine role. In certain species, such as polyandrous birds, a brightly colored female plays the "dominant" role in courtship, mates with several males and usually delegates rearing the young to their father. This behavior is commonly considered sex role reversal.

Her behavior is defined as clearly "masculine," and many scientists are reluctant to relinquish their cherished image of the male's monopoly on courtship behavior and sexual drive.

Indeed, scientists, who should be the first to point out inconsistencies and to acknowledge discoveries contrary to the accepted dogma, are caught in the same stereotypical rut as the average person. The way a biological question about gender differences is phrased reflects the gender expectations of the investigator. When an investigator classifies behavior as sex role reversal, he is stating his biases: Why should a female jacana bird be caught up in cultural role expectations when she is following the dictates of her biological heritage?

One scientist who strongly advocates the biological basis of behavior is E. O. Wilson, the patriarch of a new field known as sociobiology. In his book, *Sociobiology*, published in 1975, E. O. Wilson devotes an entire chapter to sex and society, giving endless examples of mating behavior, and he describes only the males in detail—fighting each other, strutting on a communal breeding ground—implying throughout that the only sex that battles, competes, and pursues is the male. In this chapter, all the illustrations show only the male in action, with the passive female standing by or watching from the sidelines. Wilson never questions her presence, but he ignores her as thoroughly as the many species in which males do not battle to gain access to the nearby female. They are cast aside because they do not exemplify the stereotype.

The extremes of this female minimization are reached in one paragraph in which Wilson describes typical differences between the sexes: "The males of many species are larger, showier in appearance, and more aggressive than the females. Often the two sexes differ so much as to seem to belong to different species." He then proceeds to list a variety of such unusual creatures, ranging from ants to wasps to deep-sea angler fishes. All well and good, but he fails to mention that all of these species, far from supporting the stereotype, have females larger than males.

Although his work is recent enough to take account of ample research showing evidence to the contrary, Wilson merely reiterates the outdated writings of centuries of scientists, naturalists, and philosophers who, conditioned by traditional human gender roles, examined nature from the male viewpoint and dichotomized "feminine" and "masculine" behavior. Indeed, as recently as 1982, *Copeia* (a prestigious journal) published an article on courtship behavior in which, approved by the editorial staff and secret reviewers, the author blatantly reaffirmed the male imperative and embraced the masculine stereotype, ignoring all other aspects, that "female insemination is the goal of all courtship behavior." The female is often viewed as a passive sperm receptacle. Only the male woos, and courtship is primarily the masculine prerogative.

But since gender expectations are culturally endowed, we must forgive our fellow scientists their transgressions and their obliviousness to such unconscious biases. Stereotyped sex roles are so deeply indoctrinated in us that we accept the stereotype of distinct "feminine" and "masculine" behaviors as immutable, referring to any divergences from this norm as aberrances.

The stereotype of the female who is meek, not given to action, drably colored, willingly and passively accepting the advances of the male during courtship and putting her efforts into rearing a small number of young, was probably instilled into our thinking by a subconscious parallel created between female and egg, male and sperm: Egg equals female and sperm equals male. To illustrate how interchangeable the nouns can be, we cite *Sex, Evolution and Behavior*, published in 1978 by Martin Daley and Margo Wilson. In this book, which typifies current thinking, one chapter is even entitled " The Reluctant Female and the Ardent Male." Several paragraphs, as in our example below, are equally clear if the genders *female* and *male* are substituted for their respective gametes, *ovum* and *sperm*, and vice versa:

"A sperm [male] is an entity with a mission — search, find, fertilize. In intense competition with other sperms [males] of similar ambitions, it has become stripped down and streamlined.... The prize for which the sperms [males] compete is... the ovum [female] which does not move to welcome the victor but sluggishly awaits him.... The female [ovum] provides the raw materials for the early differentiation and growth of their progeny." (Perhaps the authors substituted "female" for "ovum" in their own writing without being aware of what they had done.)

Such prose echoes that of most other scientists who ascribe the difference between the sexes to the difference between the gametes:

"Different optimal strategies [in reproduction] result from the gamete size difference." We find the same sentiment expressed in another textbook on sex and behavior, *Sexual Strategy* by Tim Halliday. "The fundamental difference between the sexes [is] the difference between eggs and sperm. [It] has had a profound effect on the evolution of sexual behaviour, and males and females have evolved very different and often conflicting strategies for reproduction."

The traditional stereotypes that are embraced as feminine and masculine traits, accordingly, are rooted in the functional and anatomical differences of the sexual cells. Eggs, few in number, passive in movement, are heavy with nurturing chemicals, whereas sperm, produced by the millions, are aggressive, strong swimmers and stripped bare of nutritional encumbrances that would delay them in their competitive race to reach the egg.

Adult males, like sperm, must compete among themselves for females. Stereotypically, males are seen to invest their aggressive energies solely in strategies to subdue rivals and to win females.

The adult female is stereotyped as meek, not given to action, an observer who passively accepts the advances of the male during courtship and then puts all her energy into rearing and nurturing her young.

This kind of thinking vastly oversimplifies the evolutionary imagination. True, females make eggs and males make sperm — for after all, that is how the two sexes are defined — but beyond this difference there are no universal axioms.

Liberated Ladies of the Animal Kingdom

If we were to look at female behavior we would find that the female stereotype, like a huge snowman in the sun, grows soft around the edges and loses its recognizable shape as it melts into a blob. There are, of course, females of many species who do show such traditional behaviors, because these roles suit their life-styles. But females do not keep tradition if it doesn't work for them; species must adjust to environmental demands, and evolution has provided them with the mechanisms to meet the needs of their ecology. If a particular life-style requires that they grow bigger than males, they do. If they must be showier, they are; if they must be aggressive, they are; and if they

SPECIES SURVIVAL

decide to abandon maternalism altogether for the good of their offspring, they do. Some examples follow.

In a number of species, the female and male look exactly alike and, except for gametes, are physiologically identical. In this way, many species of clams, oysters, starfish, sea urchins, snails, and fishes do not fit the stereotypic mandate. We would defy you to say, for instance, that the oyster you just gulped down was a female or a male. Females of these species respond to cues from a biological alarm clock, spawn and abandon their eggs to the sea, uncaring, indifferent to the fate of their offspring; they are not at all "maternal."

Furthermore, females are commonly considered to be the passive recipients of the courting male, showing no initiative themselves. Yet among many seabirds, such as the grebe, the female mirrors the male in appearance *and* in courting behavior, an equal participant in a courtship ritual wherein she bobs about, raises her wings, crosses her bill with the male's, bows, stretches, and vocalizes noisily. And in parenthood, the male mirrors the female in performing chores of rearing the young. The male pelican also looks and acts just like the female. Huge-bodied, heavy-billed, comical-looking birds, they share domestic bliss on an equal time basis. They both sit on the nest and fish with equal passion to provide whitebait, smelt, pilchards, and other piscine delights for the hungry nestlings who constantly pester their parents for food.

Another stereotype—that the male more eagerly pursues the female in courtship and that the female is reluctant and coy—is exploded by the experimental results of research on diverse species. For example, a female rat in heat defies greater dangers to get to a male than he does to get to her. In the laboratory, she will press bars more often than he and disregard torturous stinging electric charges on her toes to satisfy her sexual appetites.

Among many birds, territory acquisition is assumed to be the male's job. Yet female phalaropes and Jesus birds—flashily colored, attractively attired marsh birds—patrol their own large land holdings. A female will allow several males to set up housekeeping within her estate, and, in exchange for this space, to mate with her and take care of the kids. These liberated females leave their eggs behind in the male's nest while he, drab and inconspicuous, is unnoticed by egg-eating predators. Likewise, the females of some tropical marine fishes come in electric blues, dazzling yellows, brilliant greens, and raucous reds, embellished with polka dots and stripes. Their bright colors fit in well with life on a coral reef and are irrelevant to gender.

Even size differences do not conform to expectations. Bigger size is not always the exclusive purview of masculinity, although males usually hold the size advantage among mammals. The size ratio is modest, except within the fevered imagination of moviemakers, where the towering King Kong clutches the cowering Fay Wray in his huge fist — he epitomizing brawn, power, and macho sex appeal; she embodying fragility, subordination, and cheesecake sex appeal.

Yet in some mammalian species, females grow larger than their consorts. A strange assortment, these include rabbits, hamsters, baleen whales, and bats, a mixed bag of females seeming to have little in common beyond their superior size. (Here is a point for the *Guinness Book of World Records*: The largest animal that now lives or ever lived is the female blue whale, the giant of the oceans.) For one group, the bats, there is a simple explanation for the female's superior size, related to the trait that sets bats apart from other mammals: the need to fly while burdened with the extra weight of a fetus.

When females grow bigger than males, they can carry it to an extreme. Among spiders, for example, it is common for a female to dwarf her mate, measuring perhaps ten times his length and one thousand times his weight.

The deep-sea angler fish carries a fringed bait on top of her head that mimics a piece of rotting fish — hence her name. This attracts curious small fish who are oblivious to the huge dangerous jaws just below. When these gullible prey are within striking range, the female opens her cavernous mouth, sucking the tiny Jonahs into her whalelike stomach.

This female is a loner who cruises the black depths of the ocean, and chance encounters with other members of her species are rare. When the urge comes to reproduce, there are no singles bars, no Club Med — indeed, no place to meet a male at mating time. What kind of reproductive strategy is open to her? She solves this problem by simply acquiring males at any time and forming a permanent attachment to them. And what an attachment. The male, less than a tenth of her length, bites into her skin with powerful jaws and locks himself permanently in place. Then he proceeds to shed his entire body, with

SPECIES SURVIVAL 17

one exception, the essential part: a bag of testis. Thus, the female becomes a totally self-contained unit and, as a pioneer in the deep sea, she carries with her all the tools of survival and the provisions for the future: a bag of sperm as well as huge jaws and a tantalizing decoy to counteract the paucity of food in her environment.

The females who are the largest in the world compared to their male counterparts belong to the marine echiuroid worm *Bonellia*. Females of this European species are one thousand times the length of the male and over a million times his weight. Baby *Bonellias* all start out alike: neutral, neither male nor female, they are microscopic, free-swimming larvae. When the larvae cease their swimming stage and settle down, the paths of the two sexes diverge. If a larva settles on a rocky bottom, it creeps into a crevice and grows into a worm having a plum-shaped body and a very long, thin proboscis with two leaf-shaped flaps at the end — a female. If a larva settles on the proboscis of a female, however, it remains minute and becomes a male, welcomed into the female's gut, feeding on her food, and fertilizing her eggs. She, like the angler fish, is the family's provider. Her family is a harem of males who live in her gut. Because she's stuck in a rocky fortress and cannot get out to meet the opposite sex, her strategy is to engulf them and keep them to serve her sexual appetite.

The female hyena has burst out of the stereotypical mold in her own way — by acquiring the purely masculine anatomical trait of external genitals. She adorns herself with a penis and scrotum. These females look so malelike that an apparent absence of female hyenas puzzled observers and scientists for many decades. Only by dissection did they discover the hyena's secret femaleness: an ovary and a uterus. The penis and scrotum were shams, stuffed with connective tissue. Along with the discovery of sham genitalia came the discovery that one could tell the sexes apart by size rather than by the usual method: females were the bigger animals.

The female's sham penis and scrotum have no reproductive function. They are apparently greeting organs, used as a sort of handshake. When two of these ugly beasts cross paths, they carefully sniff each other's "genitals," looking for whatever information a hyena looks for, perhaps the other animal's sex and rank in the social group.

So what remains of the feminine stereotype? She is not always drab, meek, passive. In some species, she is brawny, big, colorful, sports genitalia, and pursues the opposite sex without giving up her femaleness; she retains the capacity to produce eggs. All such "sex-role reversals" are normal, merely vehicles to carry out the reproductive mandate. She adjusts her behavior, appearance, and, if necessary, physiology to meet the demands of survival in the context of her lifestyle, as we shall see in ensuing chapters.

2

WHERE FEMALES END AND MALES BEGIN

Birds do it,
Bees do it,
Even educated fleas do it.
Let's do it—
Let's fall in love.
Electric eels, I might add, do it
Though it shocks 'em I know.

COLE PORTER's slyly suggestive lyrics ring of biological truth. Without doubt, sexual reproduction is the most popular invention of life on earth, adopted as a routine part of life-style by hundreds of thousands of species of plants and animals. The drive to mate creates a veritable orgy of billing and cooing birds, flashing fireflies, soaring eagles, entwining earthworms, and blooming flowers, all singlemindedly pursuing the opposite sex. Our own lives are enhanced, enriched, and also made singularly complicated by the necessities of copulation and its euphemistic alter ego, love. Sexual reproduction is so pervasive and of such great significance that no one ever asks a fundamental question: Why are there two sexes?

In fact, reproduction without mating—asexual reproduction—is fast, simple, and efficient. Many plants, a few animals, and most unicellular creatures do very well by cloning themselves. Every paramecium is able to give rise to new paramecia without a collaborator and whenever it is "in the mood." However, self-duplication, or cloning, as ego-satisfying as it may be, is not always the best way to reproduce. To appreciate the limits of asexual reproduction and the benefits of sexual reproduction, we need only follow the life stories of some species that produce progeny by both asexual and sexual means, depending on the weather and where the species finds itself.

These unusual critters are reproductive switch-hitters, batting from either side of the reproductive plate, depending on the curves that life throws at them. But they are useful as more than just an oddity, for they can show us the relative advantages of sexual and asexual reproduction. If an organism can choose to use either one method or the other, depending on its needs, then it should use cloning when the environment favors that mode of reproduction, and switch to sex when conditions demand it. The pattern shown by every switch-hitter reveals that, simply stated, cloning is the best strategy to follow in a beneficent world, but when the going gets tough, the tough get sexy.

Some fantastically successful species follow a switch-hitting life style, and probably owe their success to it. One such creature is a tiny single-celled parasite named *Plasmodium*. *Plasmodium* straddles the boundary between sexual and asexual reproduction, using both mating and cloning to the fullest. It uses these two methods so efficiently that it is one of the unconquered scourges of humankind. Its story follows.

In the deepening dusk of a West African summer evening, a female *Anopheles* mosquito settles lightly upon the shoulders of an Ibo tribesman. Unnoticed, she pierces his skin with her needle-fine stylet. A little of her saliva spurts into the man's bloodstream to prevent his blood from clotting as she sucks it into her stomach. The saliva, ordinarily just an irritant, is now a dangerous droplet in his body, for it is alive with the slim, wiggling cells of *Plasmodium* protozoans, the cause of malaria.

After the wiggling cells have been injected into the blood, they are carried by the blood until they reach a resting spot, such as the liver. There *Plasmodium* begins a quiet takeover, actively burrowing into liver cells. It quickly begins growing and dividing in this comfortable home. After a while, the parasitic burden kills the liver cells, which break open to release hundreds of cloned replicas of the initial invader. Some of these replicas infect neighboring liver cells, turning out thousands more of the microscopic killers. Others work their way into blood vessels, where they invade red blood cells. Once again, the parasite grows and divides rapidly, feeding off the abundance of the red blood cells, requisitioning nutrients intended to keep the cell healthy. Soon the weakened corpuscles break open and, like their fellow victims, the liver cells, release a horde of identical parasites, each one ready to feed on other corpuscles.

The fever and chills of malaria are the outward signs of thousands of red blood cells dying all at once, of millions of parasites simultaneously breaking out into the victim's bloodstream. In the mosquito's saliva, a few insignificant protozoa, altogether smaller than the dot on this i, can within a few days become so numerous within the human body that they prostrate an adult with a serious, often fatal, disease. The feverish pitch at which the disease progresses is attributed to the power of asexual division.

Successful as the protozoan is in a single person, to the horror and grief of family and friends, the pest must find new hosts to infect. Should the tribesman die of malaria, or yellow fever, or old age, all of the *Plasmodia* will die, too. Sooner or later they must move on. Having infected one human, they must hitch a ride on another mosquito, leaving the reliable safe harbor of the human in search of another. *Plasmodium*'s trip through the gut of the insect and into a new warm-blooded host — perhaps a monkey, perhaps a man — is fraught with danger and uncertainties, and its destination is unknown. It must be ready to face any adversity. In this uncertain environment sexual reproduction holds the advantage.

Readying themselves for their trip into the unknown, some of the parasites in the corpuscles metamorphose into two kinds of cells, which deform the blood cell but do not break it. When these deformed cells are sucked into the stomach of a hungry mosquito, the parasites break out of the corpuscle and reveal themselves to be two kinds of gametes — the egglike large gamete and the spermlike wiggling gamete. There, within the moist, dark confines of a mosquito's gut, they unite, forming a sexually created *Plasmodium* with a new mélange of genes. Then they worm their way into the cells of the mosquito gut, where the individual parasites divide asexually and finally work their way to the salivary gland, ready to start the insidious cycle again.

Another switch-hitter lives half a world away, in a quiet small pond. A tiny planktonic animal called *Daphnia* (the water flea), it is as different from *Plasmodium* as day is from night. A diminutive relative of the crabs and shrimp, *Daphnia* is a harmless vegetarian, rather than a fearsome single-celled parasite. But in their reproductive habits, *Daphnia* and *Plasmodium* are kindred souls.

Peculiar-looking, *Daphnia*'s microscopic body bulges in a teardrop shape, and its two long appendages sweep the water jerkily, making it look to the imaginative observer like its voracious bloodsucking namesake, the flea. But it is an innocuous creature, feeding on algae, not on blood, and is itself, in turn, a favored food of small fishes. *Daphnia* is successful, widespread, and abundant in lakes around the world. And almost all of these *Daphnia*, from Connecticut to Tibet to Poland, are females.

In the spring, as warmer days melt the last thin panes of ice covering the lake, female *Daphnia* hatch out from hard-shelled eggs buried in the muddy bottom. Nibbling at the burgeoning algae population of spring, those females grow rapidly and reproduce rapidly — asexually. They need no male to fertilize their eggs, for these eggs can develop without sperm. Only a few days apart in age, their daughters, granddaughters, and great-granddaughters are all clones, continuing to reproduce asexually during the bountiful, algae-filled months of summer.

As the year wears on, the lake becomes exhausted of some of the nutrients needed by the water fleas. Essential chemicals become scarce. Finally, the impoverished environment throws a biological switch in the *Daphnia*, who respond to the oncoming deprivation by producing sons as well as daughters.

Exactly how the changed lake waters stimulate the females to produce males is not known, but its benefits are clear. Ten generations of female *Daphnia* producing identical daughters are ideal for life in a nuturant and warm environment, but the hardships and uncertainties of the coming winter demand new gene combinations. Only sexual mating can provide that. Indeed, the mating of female and male water fleas results in a special kind of daughter: an egg wrapped in a particularly thick shell and imbued with the ability to suspend development, to stop its metabolism and rest until the winter passes and spring comes again. Only eggs produced by a sexual union survive to hatch in the warming, welcoming lake and start the asexual cycle again.

Many others from all kingdoms of life share the split personality of the reproductive switch-hitter. Insect pests, the aphids or plant lice, whose wingless green bodies quickly cover the stems of roses, don't all crawl or hop to that stem; they hatch there from parthenogenetic eggs. They reproduce so rapidly that their natural enemies, ladybugs and gardeners, can scarcely keep them under control, and this unexcelled ability to increase their numbers is endowed by cloning. But autumn brings an end to summer's bounty, and with autumn —in a change echoing that of the water flea—two sexes appear. Winged male and female aphids mate and fly off to lay fertilized eggs awaiting the vernal growth of next spring.

Among other switch-hitters, tenacious Bermuda grass infiltrates asexual runners into every corner of the garden and wafts its sexual seeds to the yard across the street. Tiny rotifers, elegant members of the plankton, grow females during the high season and males when the pond runs out of groceries, again remarkably paralleling the lifestyle of *Daphnia*.

Sex: Playing to Win

Although switch-hitters include both plants and animals, terrestrial and aquatic, parasitic and predatory, they all use sex and cloning in the same manner. Predictably, cloning is the chosen style when food and space are bountiful, and sexual union comes into play when the animals are faced with uncertain and often deteriorating conditions. Asexual reproduction, fast and efficient, permits organisms to make the most of a hospitable environment. Sexual reproduction, the mother of variation, permits them to survive the unknown dangers of changing seasons or untested habitats.

It would seem that switch-hitting has everything going for it — an ideal system to make the most of good times and to combat the hardships of bad times. But most species have not adopted it as a reproductive strategy.

In order to be a successful switch-hitting species, each generation must have only a very short time on earth. If you live only for a week, your species can squeeze many generations into a single summer, and the environment will change but minimally during your week on earth. Your experience on earth approaches hedonism; you can live for the day and not worry about having to deal with the hardships of winter, still months away.

Unfortunately, switch-hitting is a strategy that does not suit long-living animals like humans who live throughout the heat of summer, the cold of winter, feast or famine, good and bad times.

Why bother with sexual union? Very simply, it provides genetic variation. Each and every sexually produced offspring is genetically unique, is an individual who differs from all other living creatures. In contrast, the offspring of an asexual parent are exactly like each other and the parent; generation after generation—through children, grandchildren, great-great-grandchildren, and so on—there is no genetic variation in the line. But every step of sexual reproduction is designed to generate variation, and this is the triumph of sexuality over its apparent shortcomings and inconvenience.

Living things face a world of uncertainty, a game of chance as well as skill. The object of the game is to win the bit of immortality ensured by successful reproduction, but that goal is threatened by an endless succession of hurdles and tribulations; by the possibility of being eaten in one's youth; by starving during droughts and drowning in floods; by freezing during late springs and early winters; and by the uncomforting reality that more skillful competitors are playing the same game. Asexuals play the game by finding what seems to be a winning combination and sticking to it play after play. Since the progenitor is a winner, chances are the offspring will be, too. Crabgrass spreading through a summer lawn and bread mold covering a stale loaf both demonstrate the powerful strategy of the asexual to completely take over a suitable habitat in a very short time.

But when conditions change, asexuality isn't such a good strategy. Asexuals can't make rapid changes in their genes. They cannot deal with an altered situation, and they die out just as rapidly as they proliferated.

To exemplify the pluses of genetic variation, consider bacteria multiplying in the nurturing environment of an agar culture. Each individual bacterium gives rise to a clone of millions of cells in a matter of days. Should the environment deteriorate (for the germs) by the introduction of a good dose of penicillin, reproduction ceases, and most of the bacteria die. Only one in a million, which had previously mutated in such a way that it resists penicillin, is able to grow and divide. Dose the culture with streptomycin instead, and a few different bacteria survive to reproduce in the same way. But if both penicillin and streptomycin are mixed into the culture, the ultimate bacterial catastrophe occurs; not a single germ has the one-in-a-million-millions chance of being resistant to both drugs, and every germ gets wiped out. Although one or a few bacteria had a mutation to resist penicillin, and one or a few other bacteria had the mutation to resist streptomycin, so far no bacterium had mutated to resist both. The only way an asexual species can acquire both resistances is to wait for the second mutation to occur in an individual already sporting the first. Time runs out before this might occur.

Sexual species don't have to wait. If the bacteria mated, there would be a good chance that a "mother" with a penicillin-resistant gene would mate with a "father" having a streptomycin-resistant gene. The new germ would be endowed with both and be a Noah making it through the deluge of antibiotics.

This potential for quickly and successfully dealing with change — the ability to make new gene combinations out of the existing genes of two parents — is the benefit bestowed by sexual union.

Sexual union is the original genetic engineer. It invented, perhaps a billion years ago, what humans have just begun to discover: how to take genes from different individuals and recombine them in totally new ways. Human beings experimenting on genetic engineering try weird combinations. She makes bacteria a little bit human by inserting a gene for making our own brand of insulin; he tries to rival Believe-It-or-Not by urging a tomato plant to grow potatoes on its roots. Sexual fusion is a little more conservative in its choice of genetic donors; it restricts its experimental crosses to one species at a time. But the processes, and the results, are fundamentally the same. Sexual union takes the genes of two separate individuals, extracts them into the flasks of the gametes, and experimentally recombines them in the test tube of the newly fertilized egg, the zygote. Each sexually produced offspring is a true evolutionary experiment, because the process of gamete fusion creates individuality. And no two experiments are the same, because in all probability, the genes will never extract and recombine in just the same way, so no two zygotes get an identical set of genes.

The offspring are uniquely different from the parents in unexpected and unpredictable ways. Many of the experiments of sex (like those made in a laboratory) will be, frankly, duds: offspring whose new genetic combination just isn't up to the standard of either parent, like a tomato-potato cross lacking both potatoes below *and* tomatoes above. But others will be glittering successes: the Galileos, Curies, Einsteins, Mozarts, Beethovens, Michelangelos, Leonardos of this world.

Once sexual reproduction took hold, it seems to have been only a small step to the evolution of the two sexes. But like so much else in biology, the steps of this pathway are shrouded in mystery.

In many single-celled organisms, sexual union occurs between individuals who differ a tiny bit in their chemical make-up but not at all in anatomy, nor kind of sex cells. Matings could and probably do occur between identicals composed of the same stuff.

Many single-celled protozoans are apparently distinguishable as different sexes only because they "smell" differently to one another. Matings between these tiny aquatic organisms, stimulated by the smell of another "sex" (in some species, as many as eleven opposite "sexes"), produce sex cells commonly called gametes that are absolutely identical to one another in size, shape, and content, and which fuse to produce a new individual. But identical gametes are restricted to these microscopic, simple water-dwelling organisms who are constantly crossing paths in their ponds.

The invention of anatomically different sex cells seems to have been evolution's better mousetrap. Clearly, the evolutionary world has beaten a path in that direction, for anatomically different sex cells, eggs and sperm, are found in mammals, birds, lizards, frogs, insects, spiders, lobsters, shrimp, squid, snails, starfish, worms. You name it—it has two sexes with two kinds of gametes.

Although the nuclei are similar, the external appearance of female and male gametes are vastly different. The male gamete, the sperm, is built to swim; it is energetically cheap for the body to make, tiny, sleek, nothing more than a dense packet of hereditary materials driven forward by the side-to-side lash of its whiplike tail. The female gamete, the egg, is designed to nurture, energetically expensive to make, its hereditary materials surrounded by many foodstuffs. Why has evolution opted for such vastly different gametes? Theoretically, both female and male gametes could be mobile and spermlike or large, nurturing, and egglike. But it would not be an efficient way to make a zygote, the first step of creating a new organism. For example, two sperm merging would not have enough food reserves to supply the energy a zygote needs to divide its cell to form an embryonic precursor. Two eggs, on the other hand, would have plenty of food, but they would never be able to find one another, because, being so large and heavy, and without a means of locomotion, they couldn't move.

Perhaps in the dim distant past, gametes were all alike, somewhat resembling the gametes of one-celled animals today. Presumably, ancestral gametes were jacks-of-all-trades, looked the same, were intermediate in size, barely capable of moving about, with small amounts of yolk. If these gametes were the product of some archaic oyster, during spawning season each oyster would expel thousands of such sluggish gametes. Although oysters live crowded shell to shell, a few millimeters' separation might be insurmountable for these similarly functioning sex cells. Heavy and sluggish, most would be doomed even before they had gotten started, because chance encounters of making a zygote would be discouragingly low.

The story changes with two specialists, each a master at its trade. The tiny cell, the sperm, can be produced by the millions and is capable of swimming over relatively long distances, compared to its size. The lack of motility in an egg becomes irrelevant, since many swimming sperm can find it and completely surround it.

A Sperm Is a Sperm, But an Egg Is — ?

Once the functions diverged, one gamete was freed to develop into a variety of forms and shapes, while the other, constrained by its function, became conservative in design. That conservative gamete belongs to the male, limited by its essential need to swim. Hence, it must always be small, in a fluid medium, and have a tail. So conservative is its anatomy as a result of these functional constraints that evolution has kept sperm looking alike among virtually all species. If we compare an elephant sperm, an ostrich sperm, a hummingbird sperm, a turtle sperm, an octopus sperm, and a sea-urchin sperm, the differences are so minor that it would take an expert eye to distinguish among them.

The nature of sperm has restricted the need for evolutionary options because, to parody Gertrude Stein, a sperm is a sperm is a sperm. Reproduction places only one simple demand on the sperm: it must be able to swim to the egg. As a result of this relatively simple demand, in sexual reproduction the male, the sperm-bearer, often becomes only part of the constellation of events surrounding the fertilization of an egg.

Eggs, on the other hand, immobile and heavy with nutrients, are free to differ in size, be shrouded in different kinds of skins, and contain different amounts of nutrients according to the nourishment needed by the developing embryo. Egg size, often independent of the size of the mother, ranges from the naked microscopic egg of an elephant or sea urchin to the tiny shelled egg of a hummingbird to the jelly-covered egg of an octopus to the leather-coated medium-size egg of a turtle to the huge, hard-shelled egg of an ostrich.

Despite this vast range of sizes and shapes, each egg contains the same amount of hereditary materials as the sperm of its species. All of the egg's variations depend not on the demands of fertilization, but rather on female strategy. The constraints of the female's life-style determine how much yolk she must endow to her eggs, how she must protect their delicate interiors, where she must place them to insure fertilization, to guarantee the birth of the next generation.

We should hold gametes in awe, for packed into the nucleus of each are the genetic codes to make each species' next generation. Yet potent as they are, they are not in a sense very "healthy" cells. A gamete can maintain itself and survive briefly, but in a kind of cellular limbo; it cannot divide because it has only half the number of species chromosomes. Unless it fuses with another gamete, it will die. Fusion of gametic nuclei from egg and sperm is fertilization; the bottom line of sexual reproduction is this fusion and the concomitant creation of a zygote, a cell genetically completed with pairs of matched chromosomes.

At fertilization, a single sperm leaves its filamentous tail outside and plunges headlong into an egg. The sperm, bearer of the hereditary materials in the form of a chromosome bundle, starts a biochemical reaction in which, through cellular pulsing and pushing, its bundle is forced through the egg toward the chromosomes waiting within the egg's nucleus. In the process of matching up egg and sperm chromosomes, the sex of the new organism is determined.

A human being has twenty-three pairs of chromosomes. In twenty-two pairs, both chromosomes are the same size with the same number of genes. It is the twenty-third pair that determine the individual's sex. Since sex determination in humans typifies that of all other mammals, and since our concern is strongly chauvinistic toward our species, we shall use the human as our prototype here.

The two sex-chromosomes types are simply identified as X and Y. Women and other female mammals have two matched X chromosomes as their sexual pair; men and other male mammals have a mismatched set, containing an X and a Y. Looking at the differences in anatomy, physiology, and behavior between the sexes of most mammalian species, we might naturally, but incorrectly, assume that these dichotomies reflect sizable genetic differences as well. But actually, 98 percent of the chromosomes direct the making of a human being and only 2 percent guide the determination of sex. This tiny Y is the only genetic difference between females and males.

The disparate sizes of the X and Y reflect their genetic content. Y genes are sparse while the X chromosome, five times the size of the Y, is chock full of hundreds of genes, many vitally important, and all of them apparently missing on the Y. For example, in humans the X has genes to control the digestion of sugar, to direct blood clotting, to determine ability to see light and colors. The poor male is in a precarious position with regard to all these essential biochemical directives. If he should have the bad luck to inherit a defective gene, he's stuck with the characteristic it carries. He has only one copy of the X and hence only one set of its genes.

The female's double X gives her a large dollop of immunity to mutations on this chromosome, for even if one partner of the gene is defective, the other is almost always normal. Thus the male shows the effect of X-chromosome mutations much more often than does the female, and this unequal susceptibility is a cross men must bear. Human X-gene mutations suffer men to fall prey to such disabilities as hemophilia (bleeder's disease) and red-green color-blindness, as well as an incredibly long list of other six-linked difficulties including congenital night blindness, congenital deafness, juvenile glaucoma, Hurler's syndrome, and spondylo-epiphyseal dysplasia. More than 8 percent of American men suffer from some degree of red-green color-blindness, as opposed to less than 1 percent of American women. Men with one color-blindness gene are indeed color-blind, but women with one color-blindness gene are just carriers: unaffected themselves because of the normal gene they also have, but fully capable of passing on the defective one to their sons, who would then be color-blind.

With his "odd couple" of chromosomes, the mismatched X and Y, the male mammal is a genetic anomaly, but the female has an oddball trick up her sleeve, too. Very early in her embryonic life, soon after she receives her two X's, and at a time when all the chromosomes should be working away like crazy, intent on turning a creature of a few hundred cells into one of a few billion, an extraordinary thing happens: In almost every cell, one of her X chromosomes just curls up and goes to sleep. As if in sympathy with the deprived male, the female relinquishes the use of one X chromosome and makes do with the other. Only cells that retain two active X chromosomes are destined to become eggs.

Even if the female uses only one of her X's in each cell, she doesn't fall heir to the difficulties that plague the male. Although one X chromosome dozes in each of the female's cells, each cell makes up its own mind about which X to put to sleep. One cell may use an X with the gene for hemophilia, for example, but the woman won't be a bleeder, because a neighboring cell will use the other X and turn out lots of clotting factor.

Called the Barr body, the dozing X is a landmark in the female cell. Curled up in a corner of the nucleus, it is a tight little bundle of DNA, a microscopic badge of femininity. Used as such, doctors can tell the sex of a sixteen-week-old fetus by extracting and analyzing a few cells. Although this process — known as amniocentesis — is used to hunt for genetic problems, an extra dividend is the determination of the unborn child's sex.

Because of her Barr bodies, the female mammal is a genetic mosaic, made up of two kinds of cells: those with active X #1 and those with X #2, much as a mosaic might be made up of blue and green tiles. This additional fillip of genetic diversity means that identical twin girls aren't all that identical, because their "blue and green tiles" are arranged differently. Indeed, female identical twins tend to look less alike than their male counterparts.

Still other oddities of inheritance arise from the fact that the X chromosome has more genes on it than the Y. Since a mother endows both sons and daughters with X chromosomes, but a father bequeaths an X chromosome to his daughter and a Y chromosome to his son, we confront the paradox that, though they both receive the same number of maternal genes, a girl gets more paternal genes than her brother, and a boy gets more maternal than paternal genes. Could this account for the familiar observation that boys often seem to resemble their mothers while girls look like their fathers?

The Y does have a few genes, although rather peculiar ones — genes that determine hair in the ears, "porcupine" skin, or webbing between the toes — traits seemingly left over from our evolutionary ancestry. But along with these weird visible traits (or phenotypes), as we shall see later, the Y carries genes that direct the production of sperm — bizarre cells bearing flagella, looking not like mammalian cells but more like one-celled protozoa.

Female or male gender, despite its determination at the moment of fertilization by the sex chromosomes, doesn't become apparent until the organism has taken a recognizable form. At first, the sex chromosomes, whether XX or XY, seem irrelevant to the developing embryo. The zygote divides into two, then four, then sixteen cells, and so on, until it becomes a ball of hundreds of cells, a blastula soon to become a gastrula, which, among vertebrates, is the first clearly defined form of a creature with an emerging nervous system. All this takes no more than a few days, and by the time a woman discovers her pregnancy about one month after conception, her offspring, attached to the uterine wall by a slender umbilical lifeline, already has a mammalian face. Afloat in her uterine warm mineral bath, only a little longer than an inch, the tiny embryo has a formed, rounded head, almond-shaped eye bulges, and budding fingers. Nevertheless, it is physiologically and anatomically ambiguous; there is no way of knowing if it will become a female, a male, or, dreadful to consider, remain a hermaphrodite.

At this stage of development, the body is not committed to anatomical statements of its gender. It is prepared to become either sex, sporting a gonad that as yet is neither ovary nor testis, and harboring genital ducts, tubercles and clefts that just as easily become either the uterus, clitoris and vagina of a female or the vas deferens, penis and scrotal suture of a male. The brain, too, is still androgynous and, as we shall see later, it will become part of the reproductive system when sex behaviors are etched into its neurons.

Within the next month or so, the embryo embarks on an irreversible path.

The choosing of female or male reproductive system is controlled by the presence or absence of that single chromosome, the Y. To choose is to exclude one of the two sexes; the embryo destroys or alters the reproductive accoutrements belonging to the other sex. Once given, the orders are not rescinded. There is no turning back, and no switching of sexes.

The puny Y comes onstage when the embryo is about seven weeks old. Into this world of sexual ambivalence, the Y makes its entrance, producing a chemical, the *H-Y* antigen, which prods the undifferentiated gonad into becoming a testis. This antigen is a product of the genetic message found only on the Y, and hence, because a Y chromosome and femaleness are mutually exclusive, the Y-bearing embryo inevitably becomes a male. With this single thrust, at the right moment, the genetic determination of sex is over and hormones oversee further sexual development.

The newly differentiated testis produces male sex hormones, the androgens. The most familiar androgen is testosterone. Over the next several weeks, testicular hormones dismantle the potentially female duct system; urge the male duct system to become a special channel, the vas deferens, for carrying sperm; and enlarge the genital bud so that a penis presides and a scrotum forms. Maleness of the reproductive organs is established.

Females develop only in the *absence* of the Y chromosome and its *H-Y* antigen. The female's duct system expands and enlarges into structures to carry eggs and to house embryos. The gonad changes into an ovary, which will manufacture eggs to become the next generation. In anticipation of the species' future, and to minimize risk and potential danger, a female is born with all the eggs she will ever shed in adulthood, each waiting in her infant ovary for its moment of release many years later. Hormones, too, conduct the anatomical development of reproductive organs, but since a female embryo is bathed in her mother's amniotic fluid, supersaturated with female sex hormones, it is thought that she need only passively accept the right hormones produced by her mother, while the male embryo must counteract them with his androgens.

Gender: Sex Is All in Your Head

The mandates for distinguishing specific female from male reproductive behavior obviously must occur in the brain, since behavior is controlled by its directives. Ergo, that the brain acquires the tendency to respond in sexually appropriate ways should not come as a surprise and should not be treated with any more significance than the fact that an ovary produces eggs. The ovary and uterus need the brain to direct them. Wouldn't evolution have been foolish to create such lovely and efficient systems to house and nurture an embryo and then not provide the mechanism to put the embryo there? Thus, as ovaries (or testes) are being carved out of undifferentiated gonads and their asexual channels are being shaped into fallopian tubes and a uterus (or vas deferens), small areas of the brain are becoming sensitized to react to the right sex hormones in the future. In effect, just as eggs are readied for the future, so is the brain being programmed for its reproductive functions in adulthood.

Reproductive differences are etched into a tiny primitive neural component, the hypothalamus. An oval band of gray matter, it is found at the base of the brain stem and weighs only one three-hundredth of the whole brain. During embryonic development, sex hormones, whether female or male, travel from the center of production, the gonads, to the brain, when they impinge on the appropriate brain receptor cells. At this early stage in life, hormones merely sensitize brain cells to react to the very same hormones when the individual achieves reproductive age. With such sensitization to sex hormones, the cells remain permanently responsive to the merest traces of the same type of sex hormones that visited and left a calling card during the embryo's development. If the brain cells are not sensitized early, then the adult female mammal will not reproduce.

Together with the attached pituitary gland, the female hypothalamus is scheduled in advance to direct the ovary to release eggs sparingly, at regular intervals, while the male hypothalamus is scheduled to direct the testes to manufacture multitudes of sperm on a continuous basis. In addition, working in harmony with the brain, the fetal exposure to hormones will predispose each sex to show the appropriate characteristic behavior at the time of reproduction. In the absence of testosterone, the hypothalamus prepares for female sex behavior; in contrast, under the tutelage of testosterone, the hypothalamus prepares for male sex behavior.

Thus, when the gonads become female or male, the embryonic

brain also acquires a reproductive mandate through the influence of hormones.

Since all adult human brains look alike, how did scientists find out that the brain's reproductive function is etched in the neurons long before the organism can experience socialization?

Knowledge of brain sexuality in humans comes from hundreds of studies carried out on such birds as Japanese quail, chickens, zebra finches, and pigeons, and on such mammals as rhesus monkeys, dogs, sheep, guinea pigs, hamsters, mice, and rats.

The rat, probably the most extensively studied animal, is the standard-bearer of biology and psychology, the mainstay of laboratory research. The rat has achieved fame and notoriety as an experimental animal. Prolific and easy to raise, it also has convenient patterns of early development. After a short pregnancy, the rat gives birth to pups that are born in a very immature state: hairless, helpless, and blind, fully maturing only after being nurtured awhile in the confines of the mother's nest. Such early birth makes the neonatal rat pup a perfect experimental creature, an external fetus which can be studied in ways that might require intrauterine surgery in primates. What have scientists found out about rat sexual development?

If a male rat is castrated on the first day after birth, therefore permanently lacking testosterone, he never shows male sex behavior at adulthood. Instead he becomes a behavioral female—that is, he crouches as a male mounts him. If, in addition, he is given an ovary at adulthood, he becomes a physiological female, showing typical estrus, but of course has no uterus and cannot conceive. Thus the absence of testosterone produces a female hypothalamus. Even if the neonatally castrated adult animal is bombarded with testosterone, the brain resists the change. His behavior cannot be masculinized, because his brain cells were never sensitized to testosterone at the vital early stages of development.

In the same experimental vein, a female, injected with a large dose of testosterone just after birth, will mount other females as an adult. Even with her ovary intact, she is not able to ovulate cyclically, since the hypothalamus has been programmed by testosterone to produce sperm. Her hypothalamus, sensitized to respond only to testosterone, is blind to the ovarian estrogens that float around the neural receptor cells.

These transsexual manipulations can be made for only ten days after birth. Beyond that, even massive doses of testosterone can no longer inscribe a male sexual future on the brain template. Among primates, in contrast to rodents, this critical period of sensitivity occurs long before birth. Experimentation requires intrauterine surgery, endangering the mother and generally entailing more complicated procedures. With so many variables and possibilities of mishap, experimental results are not as clear-cut, although the general pattern is the same as that shown by the rodents. The absence or presence of testosterone sets the fetal neural stage for the animal's entire sexual future.

Thus we see that just as the gonads and associated reproductive structures become female in the absence of the *H-Y* antigen and testosterone, so the mammalian brain produces eggs and sexual heat cyclically in their absence. Of course, like everything scientific, results are never completely consistent. True, the fetal brain acquires a reproductive mandate, but despite its apparent immutability, the physiological and behavioral expression of gender can be modified by environmental quirks and even by chromosomal abnormalities. Nature has been carrying out experiments of her own for hundreds of generations, corroborating the discoveries of carefully designed research and showing trends that direct future work.

Even though female mice usually show characteristic behaviors associated with estrus, some break away from tradition and include in their reproductive repertoire some behaviors that typify the male. For example, Frederick Vom Saal and F. H. Bronson, of the University of Texas at Austin, noticed that some female mice marked territories with urine, were "tomboyish," feisty, fought more than usual, and, of greatest significance, had irregular periods of sexual heat. In trying to find out why these females had developed a more "masculine" behavioral style than their sisters, these researchers backtracked all the way to the womb, making the discovery that one's wombmate can shape one's sexual destiny.

It seems that embryos in a womb are under the influence of each other's hormones; a few hormonal molecules absorbed from a neighbor of the opposite sex can cause major changes. If she is sandwiched between two males, a female fetus is directly exposed to male sex hormones, which evidently infiltrate and leave their traces in her brain and body. Even though the consequent anatomical changes in her genital tract are slight, testosterone sensitizes the brain to react to the male sex hormone (which is normally produced in small amounts by all adult female mammals). Hence, as a grown-up she is given to bouts of male-like aggression and is sexually less appealing to males than her sisters who had less contact with brothers in the womb. However, although she is in heat less often and more irregularly, she is capable of mating, bearing young, and mothering. Indeed her prowess as a mother is impressive. She becomes fiercely aggressive and battles other females for scarce resources, becoming an ardent mother whose young probably have a greater chance of survival if times get rough than do the young of a more typical female. But in good times her opportunities for mating may be far fewer than those of her more seductive sisters. Clearly an XX female's femininity suffers from her position as the middle of a prenatal ménage à trois.

Rat researchers Robert Meisel and Ingeborg Ward, of Villanova University, in Pennsylvania, not content to be upstaged by mice, conducted their own experiments. They found that since blood flows in the rat uterus from the tail toward the head, all female embryos developing "downstream" from males (regardless of *proximity*) are washed by their brothers' sex hormones. Later in life, like the similarly affected mice, these rats behaved in a masculine manner, mounting other females.

Why can't the rodents, to quote Henry Higgins, "be more like a man"? Despite this masculine manner, these females are still functionally normal. Embryo females that live in communal uteri, such as the rodents, must have a strategy to deal with the onslaught of masculinizing testosterone from their brothers' testes, an onslaught that could ruin their future reproductive success. Apparently, female embryonic mice produce an enzyme that destroys most of the testosterone before it can interfere with normal female gonadal and genital development. Perhaps very small traces get through to the brain and masculinize it just a little. Yet perhaps sensitivity to some testosterone may be a good female strategy. As we mentioned earlier, testosterone-affected females are more protective mothers and fight harder for access to food. If times are tough, such females may be winners in the game of offspring survival.

Human females and males are not immune to disruptive hormonal influence, either. Human physiology can in fact become "confused," because the adrenal glands, in addition to the gonads, also produce sex hormones, especially androgens, or testosterone. Normally, the adrenal glands make relatively minute amounts of androgens, but in some women production soars, and may even rival the amount of testosterone produced by normal testes. In a fetal male, an excess of testosterone does not affect his anatomy or physiology, but in a fetal female such overproduction results in at least partial external masculinization noticeable at birth, ranging from a slightly enlarged clitoris to a complete penis and sham scrotum. This condition is called adrenogenital syndrome, or AGS. Internally, however, the gonads and other reproductive structures are untouched, retaining their normal female functions. In addition to external masculinization, the girls often behave like boys. According to Drs. John Money of Johns Hopkins University Medical School and Anke Ehrhardt of the State University of New York, foremost researchers of human sexual development, the girls preferred the companionship of boys, played with boys' toys, engaged in boys' sports such as football, baseball and basketball and considered themselves and were thought of by others as tomboys. These remarks may seem sexist now, particularly since the acceptance of girls as pitchers and batters in Little League teams, but that was the prevailing view in the 1960s. Adrenogenital girls do not have much in common with "ordinary girls."

In these females, excess of androgens is not a result of uterine position, as it is in the mouse, but rather, it is usually the result of a genetic abnormality. In one extreme case, where a newborn girl sported a penis, she was raised as a boy, although she was genetically a female, grew to think of herself as a boy, and subsequently had to seek medical help on the brink of puberty.

In other species, some females don't fare as well as the mouse when they are exposed prenatally to the opposite sex. In species where litters are the exception rather than the rule, a male wombmate can do more than alter a sister's behavior — he can actually destroy her reproductive potential.

Cows normally have only one calf at a time, but on rare occasions they have twins. Hormones of twins of the same sex are concordant with one another, and usually each twin is normal; but if the twins are female and male, the female suffers anatomical ruin. Early in her embryonic life, she is invaded by her brother's *H-Y* antigen, which enters her bloodstream by way of the shared placenta. This female, unlike the mouse, has not acquired the evolutionary armor to protect herself from masculinizing influences. The effect of the *H-Y* antigen is overwhelming. Her barely differentiated ovary is converted into a quasi testis, which then produces male sex hormones, expands the Wolffian duct (the embryonic male urogenital tract), and shrivels the Müllerian ducts (the embryonic female urogenital tract). Curiously, the testosterone leaves her external genitalia intact, although the clitoris hypertrophies. Such a twin, called a "freemartin" by farmers, will never reproduce.

Since the cow generally sheds only one egg during her estrous cycle, a developing embryo is exposed only to its own hormones and its mother's, both of which are generally harmonious with the offspring's genes. The freemartin, a rare discordance of hormones and genes generated by twins of the opposite sex, reveals the power of hormones to imbalance the dictates of the sex chromosomes and to destroy biological harmony. Not all species are capable of producing freemartins; among most, each twin has its own placenta, and therefore an early exchange of *H-Y* antigen does not occur.

Discordance between genes and hormones in mammals—rats, mice, cattle, dogs, and humans alike—can lead to many anatomical abnormalities and, in humans, even cultural aberrances. One genetic mutation, called testicular feminization, makes the tissues of the fetus, including the brain, insensitive to testosterone. When this mutation occurs in a genetic male, he will follow a female reproductive mandate, behaving and appearing female.

In one famous case of TFM (testicular feminization males) genetic male twins were reared as girls, one to become an airline stewardess and the other a fashion model. Clearly, they were treated culturally as females even though they were genetic males and hence sterile, each with the typical anatomical characteristic of a vagina that does not lead to a uterus.

Another example of human society's ability to see femaleness in the midst of developmental confusion is Turner's syndrome. The unfortunate individuals receive only one *X* chromosome and no *Y*, at the moment of conception. They are genetically classified as *XO*, but society calls them girls, because in the absence of a *Y* chromosome and its product, *H-Y* antigen, their external genitalia and internal duct systems are female. But they are biologically neither female nor male. The lack of a second X chromosome means that, amid numerous other developmental difficulties, their ovaries cannot differentiate properly and they will never be able to produce eggs.

Differentiating the embryo into a female mammal is a directive of the chromosomes, which guide the expression of genetic sex from its determination at the moment of fertilization. The chromosomes wield great power, are difficult to dethrone, precisely programming the series of chemical events that guide the embryo toward establishing its full identity as female or male. But, as we have seen, perturbations —enzymatic, hormonal, or environmental—can occur, deterring the growing fetus from its normal sexual development. Such defects are fortunately rare.

Many nonmammalian embryos, such as those of fishes, are not under the sway of such a genetic autocracy. Among fishes, the genes that determine sex are scattered among many pairs of chromosomes. The environment in which the embryos develop is not a well-equilibrated maternal incubator but rather a variable watery milieu, and the sex of fish embryos is subject to the whims of nature.

One thorough study of sex determination in fishes was done on the Atlantic silversides, the whitebait so often used as the lure on an angler's hook. This fish normally inhabits shallow bays along the Atlantic coast, where it breeds every other week during the spring for several months. In New England, breeding begins in April, when the grassy shallows used as nurseries are still uncomfortably chilly.

Spawning side by side with the male, the female extrudes her eggs into a bath of milky white sperm. Upon fertilization, the shell hardens, and tufts of sticky threads appear, adhering the eggs to grasses and weeds around them. While attached to the vegetation, they grow into tiny fish embryos, exposed to wide variations in temperature—chilled by the night air and heated by the noon sun.

For many fishes, such vast swings in temperature would ring the death knell, but these fish seem oblivious to hot and cold, and they develop into embryos with coal-black eyes and long, wormlike tails. At the prescribed time they break out of their now-softened shells, and the newly hatched babies cluster together, seemingly, untouched by the thermal ups and downs. In fact, however, the temperature swings have already profoundly affected the future life of each embryo.

In the cold water of early spring, eggs are more likely to develop into females, regardless of their genetic sex, whereas eggs developing in the warmer water of late spring and summer are more likely to be males. Evidently, temperature profoundly affects the instruction of the genes through mechanisms that we do not yet understand. But we do appreciate the end result. Evolution gives advantages to the female, the egg producer. Fish grow throughout life, and earlier hatchers have a longer time to grow before their first breeding season. With this head start, they have a size advantage over the late seasonal hatchers, mostly males. Large size is more important to the female. The larger the female, the greater the body cavity and the greater the number of eggs that can be stored until spawning time. More eggs lead to higher fecundity and to more progeny. The male need not be very large, for he can pack millions of sperm in an area the size of one egg.

Other species, too, allow thermal influences to obliterate the genetic mandate. The green sea turtle produces females when the temperature is high and males at lower temperatures, while the alligator produces males at high temperatures and females at lower ones. (Again, scientists do not understand how this sex-determination system works.)

To the starworm, *Bonellia viridus*, neither genes nor temperature matter. One's sex and one's place in life go hand in hand. An uncommon species, the starworm is wedged, immobile, into rocky crevices along the Mediterranean coast. An adult female looks like a small balloon with a very long string attached. The balloonlike body is three and one-half inches long, but its stringlike neck may be longer than three feet. This improbable creature begins life among plankton as a microscopic larva trimmed with fringes that wave to and fro, propelling it through the water. At this stage, it is asexual — without even a gene to dictate if the larva will grow up to be female or male. *Position in life is everything* could be embroidered on the starworm's home sampler. If the larva is cast by the sea among some rocks and remains there tucked into a crevice, safe from the torturous tides and chastening currents, this sexless larva becomes a female. Once secured in its homestead, it starts to change its shape, ballooning its body and dispensing with its swimming fringes, since it will remain homebound forever. From the balloon, a long slender neck stretches toward the opening of the crevice. This proboscis may even grow as tall as a man, allowing her to stay completely hidden deep in the crevice while keeping her lips, at the tip of the neck, out on the front porch to sample the passing smorgasbord of planktonic delights.

But she doesn't eat every tiny passing creature. Among her selections are other planktonic starworm larvae being carried along by the same currents that carried her to a cozy seaside home. These larvae are neutral, neither female nor male, waiting just as she did to find their position in life — to find their karma in a rocky niche. If one of these larvae happens to land on the female's lips, its biological future is fixed, just as hers was when she found an unoccupied crevice. A chemical on the female's lips metamorphoses the neutral larva into a male, who remains barely one-eighth of an inch long. The male finds his way into the uterus and is kept by the female, functioning solely as a provider of sperm for her eggs and relying on her completely, gigolo-style, for nurturance and whatever TLC a starworm can lavish. In return for his sexual services, he gets full room and board.

The female keeps a harem of these microscopic gigolos, following the principle of "More is better," and assuring her a convenient and generous supply of sperm, ready whenever she needs them for her eggs. Keeping males stocked in her uterus overcomes the disadvantages of never being able to leave the crevice.

Sex in the starworm is decided not by sex genes, but by place: a truly efficient mating system, since no animal goes to waste. Every crevice is occupied by a female. Every neutral larva that lands on her lips is chemically transformed by her into a male. This is ecological efficiency carried to the extreme, maximizing the number of egg producers in a rocky area and optimizing the chance that all the eggs will get fertilized.

The Queen Bee and Her Genetic Monarchy

Although comparisons between ourselves and such bizarre creatures as starworms seem a bit farfetched, some qualities are shared. For instance, both humans and starworms require special chemicals to become males.

Most people are familiar with human sex determination as a chromosomal event: the chance pairing of chromosomes at conception. Yet, as we saw, gender can result from other determinants, such as position in space, chemicals, or water temperature at critical times during development.

In one group, the female is created by a combination of genes, environment, and peer pressure. Among honeybees, the queen bee is an absolute monarch. She bequeaths genes to each egg by deciding in advance what kind of eggs, female or male, she will lay. She creates daughters or sons as she sees fit, controlling the sex of her offspring by laying eggs that are fertilized or unfertilized. These choices are possible only because she stores sperm in a special sac, sperm gathered on her nuptial flights soon after she had been crowned a virgin queen. This sperm remains viable for several years after her mating adventures. To make daughters (more female workers to serve her and the hive, or future queens to found new hives), she releases sperm to fertilize her eggs and lays them in appropriate hive cells. But if the queen desires sons, she simply lays eggs that have not been fertilized. Amazingly, these unfertilized eggs become her sons, the drones who will mate with a new generation of queens. This is how bees determine sex. Males are descended solely from their mother's genes and are a rare example of healthy haploid organisms, animals with only one of the two sets of chromosomes.

But the queen bee, monarch of the hive and absolute dictator of the sex of her offspring, does not lay fertilized or unfertilized eggs solely on the basis of her regal whims. She is subordinate to the demands of the hive, actually a slave to her environment, for the kind of egg she lays depends upon the kind of brood cell she has to fill. Sterile female workers, queens, and drones grow up in waxy cells of different sizes. A queen must place male eggs in drone cells and female eggs in worker or queen cells. The worker bees decide if the hive needs workers, drones, or queens, and then build the appropriate kind of brood cells. The queen, mother of the workers, becomes their servant by fulfilling their demands and laying the right kind of egg in the brood cell. An empty drone cell is filled with the appropriate unfertilized egg while a worker-size cell is stuffed with a fertilized egg. A queen cell gets a fertilized egg, too, but a rich, highly nutritious diet of royal jelly makes this egg into a queen. It is really the hive as a whole that masterminds sex determination, but the queen bee's prognosticating ability to procreate the sex of her choice is an enviable quality, one not found among the most intelligent species.

All the examples that we have surveyed emphasize the principle that an embryo, starting out as a neutral, asexual being, never remains in that delicate, unstable position; some force inevitably pushes it in one direction or the other, committing it early in life to a male or female reproductive mandate. Once started on its path, the embryo gathers momentum, acquiring the attributes of its gender effortlessly and rapidly. Trying to stop the embryo as it accumulates the characteristics of its own gender is like trying to stop an avalanche.

Among mammals, the embryo's early asexuality does not mean that it has the potential of becoming *either* female or male, even though it houses a bisexual reproductive system. Which system will see the light of day results from a medley of interacting forces: chromosomes, hormones, and environment. XX chromosomes increase the odds of the embryo becoming a female, whereas an XY chromosome pair increase the odds of the embryo becoming a male. But chromosomes by themselves do not guarantee that an embryo will grow into a normal female or male. Fetal mammals whose sexuality is confused by hormonal or other errors are headed for a reproductive dead end. Other groups, however, have been favored by evolution to experience what humans can only fantasize about: what it's like to be the opposite sex.

Sexual bipotentiality is so vigorous among some groups (e.g., invertebrates, fishes, amphibians) that even after the embryo has hatched, acquired a sexual identity, and become a sexually functioning adult, it can change into the other sex. Such changes are the rule in the slipper shell, a common seashore snail with the provocative name of *Crepidula fornicata*, so named because more than a dozen snails, stacked on top of each other, live in permanent polygamous union. The big one on the bottom is always the female, and a male becomes a female when another slipper shell lands on top of it — ergo, its name, based on the conventional sex position. Thus, becoming a female depends on who's on top and can happen anytime in the life of a slipper shell. Like the starworm, position is critical to femaleness and, also like the starworm, the female always has a male, for he is the snail on top. The top snail remains a male until he is superseded by another, and then he converts into a female and produces eggs.

Among fishes whose gender is loosely controlled by genes, sex reversal is the way of life for a number of species, many of them residents of coral reefs, which have a social order maintained by a pecking order. Given the right social situation, a female gives up her femininity and opts to become a domineering male. By doing so, she rises to the top of the pecking order, where she acquires control and power—not over money, but over the reproduction of subordinates. It is relatively easy to convert her ovary to a testis and to acquire color patterns and behaviors of a typical dominant male, but it is not so biologically simple to change a testis into an ovary, and few fish have the option to convert into females. Some do, however. The clown fish, called that not because of behavior but because of its brilliant red and white stripes, swims with impunity amidst the waving, stinging, noxious tentacles of a sea anemone. Clown fish live socially in mated pairs, one pair to an anemone, along with several adolescents who live in schooling groups. If by chance two adult females move into the same anemone's forest of tentacles, life is full of strife. The females battle to the death, and only one survives to become the headmistress of the cluster of adolescent clown fish and of all she surveys. She, matriarchal head of the clowns, mates with the top male adolescent—that is, the one who subdued all his rivals—and he becomes her consort. Should the matriarch die, an amazing conversion takes place: Her widowed male changes into a female and steps into her slippers. The former "he" develops ovaries, acquires the female's bossy qualities, and mates with the next male in succession, who now becomes "her" prince consort. If he doesn't enjoy his newfound femaleness, he cannot revert back to being a male. One sex change to a customer.

Having It Both Ways

Other species of fish can switch their sex back and forth at will. Tiny groupers, living in the warm waters of the Gulf of Mexico, are true hermaphrodites, each fish having eggs and sperm ready at any time. However, they are cautious about when they use their gametes, and they avoid the ultimate narcissism, that of fertilizing their own eggs. Hence, at any moment they will release only eggs or only sperm. During mating, to decide who gets which gender initially, upon meeting, each pair of fish do battle—nipping, pursuing, and trying to subdue each other. The winner—usually the larger of the two—gets to be the first mama. The other fish darkens, pursues her, and spurts out sperm over her eggs.

After shedding her eggs, Mama, who still has her own sperm, switches her behavior within a matter of minutes: She darkens and pursues the other fish, still swollen with "his" eggs. He now becomes the female, behaving in expected traditional style, and he sheds these eggs as the former mama sprays them with sperm. Both fish produce eggs and sperm, alternately, each momentarily a female, then momentarily a male.

Because they never make a permanent commitment, they all keep their ovaries, testes, and associated hormones in running order and can switch to one gender or the other, depending on the traffic signals. Among these groupers, donning the appropriate behavior and appearance is just as easy as changing clothes.

In our own species, the permanence of gender dogs some of us from the moment of birth. How often has a man said, "I wish I were a woman" or a woman said, "I wish I were a man"? The discoveries of sex-reversing animals should have stimulated reams of inquiry and endless experimentation, yet few scientists research this fascinating and biologically confounding phenomenon. Perhaps it upsets our entrenched subconscious concepts of the immutability of gender; perhaps it unnerves us to contemplate unisexuality, a reproductive continuum between female and male.

Such trepidations are not experienced by writers of science-fiction fantasy. Understanding that many humans, curious beings that they are, would like to know what it would be like to be both sexes, writer Ursula K. LeGuin has created such a world in her novel *The Left Hand of Darkness*, in which each human is both sexes, functioning just like the grouper. In this imaginary world of androgynous people, "When the individual finds a partner... hormonal secretion is further stimulated... until in one partner, either a male or female, hormonal dominance is established. The genitals engorge or shrink accordingly, foreplay intensifies, and the partner, triggered by the change, takes on the other sexual role.... Normal individuals have no predisposition to either sexual role.... They do not know whether they will be the male or the female and have no choice in the matter.... No physiological habit is established and the mother of several children may be the father of several more." Although the grouper probably did not serve as her model, it could have been her guide. Thus, in this imaginary world, just as in the real world of the grouper, becoming a female is not up to the sex chromosomes but is determined by spur-of-the-moment arousal. These females are transitory and ephemeral, but they fulfill the reproductive needs of the species.

Humans should not be hidebound by their biological make-up, for they can transcend these limits culturally. Psychological gender identity — how we see ourselves as female or male — grows not out of our chromosomes, or even out of our gonads, but instead out of the way society sees us.

A baby's entry into the world is immediately greeted by a label: "It's a girl," or "It's a boy." The blue diaper pins or pink hair bow then confirm the appearance of the external genitalia. Regardless of our genes, chromosomes, and hormones, this categorization, female or male, will most likely stick with us throughout life. The testicular feminization XY "girls" who are really chromosomal males seem to be just as female as true XX girls, probably because they are raised as girls and indoctrinated with the cultural expectations of females.

The embryo comes a long way from the fusion of sperm and egg. It weathers the chromosomal imperatives, the dictates of the hormones, the buffeting of external vicissitudes. But the embryo has a long way to go before it emerges from the maze of developmental complexities as a fully functional member of its gender. The nine months that a human female spends inside her mother's uterus — from the time a sperm, bearing an X chromosome, fused with the egg, to the time the beaming father hands out pink-wrapped cigars — is merely a tiny fraction of the time that it will take to become a woman.

For humans, the road to adulthood is not simple. Once out of the womb, humans must contend with the subtle, shifting, infinitely variable tides of culture. These cultural influences — what our parents do and say, what we read or watch on TV, what our peers decree — affect a critical part of sexual development and engender the psychological awareness of what it means to be a female or male.

Human beings love to classify, to label things, and human society cannot tolerate sexual indecision. Endowing each baby at birth with the designation of *male* or *female* emphasizes the separation of the sexes. This allows gender-determined "females" like the *XO* or the *TFM* to function relatively well psychologically and socially as girls and women. Biologically they will never function as females. Only when the individual reaches puberty does her true nature as a female mammal appear, distinguishing her from all other animals. Only at puberty does the hallmark of femaleness, hormonal cycling, step to the fore.

3

Transition

Humans and other mammals don't metamorphose in the manner of insects and frogs, but they do go through their own style of metamorphosis, called puberty.

The dormant reproductive system, resting quietly for months or years as the mammal grows, finally begins to stir and stretch itself. This subtle, gradual transition slowly molds and reshapes the animal inwardly and outwardly, taking several months to remodel a filly into a mare and several years to remodel a girl into a woman.

And the stirring of this system sparks changes that, while not as outwardly dramatic as the adult cicada stepping out of its unzipped pupal skin, or a tadpole growing the legs of a frog, are equally revolutionary. Almost every part of the body is affected.

Although we are aware of the upheavals of puberty, we are oblivious to its initial triggers, to the switches thrown to begin the trek to adulthood. These triggers are a complex mixture of crucial internal and external events, many of which scientists are as yet unable to pinpoint, even in the female human.

Usually, we look for the onset of puberty at a certain age, but in girls, there is a more accurate gauge for crossing that threshold. At puberty, fat is clearly beautiful.

Puberty, like a bathroom scale, "weighs" the girl, and when she reaches approximately one hundred pounds, about one-quarter of which are fat, she pubesces, regardless of her age or height. The specific mechanisms have eluded science, but somehow the brain—probably the hypothalamus—monitors the amount of fat in the body. Estrogen, incidentally, collects in fatty tissue, so the hypothalamus may measure fat content indirectly by gauging estrogen levels. A higher level of the stored hormone would correspond to more fat.

As annoying as it may be to have all this fat in one's body when one wants to be Los Angeles skinny, it makes good biological sense. Puberty means potential motherhood, and a girl can't be a mother until she has stored enough fat to nourish a fetus through nine months of pregnancy, an energy demand of up to 80,000 calories. Lo and behold, twenty-five pounds of fat stores 87,500 calories. So if hamburgers and malteds are scarce, the female human can still carry a fetus to term, because at pubescence she has a fat savings account on which she and her fetus can live.

Within the last century, improved nutrition in the Western world has decreased the average age of puberty by two years, from about 14.5 to 12.6. Conversely, poorly nourished girls, such as ones suffering from economic deprivation or anorexia nervosa, pubesce later, because they reach the critical weight later. Some may never reach it at all. The current crisis in teenage pregnancy may be at least partly due to improved nutrition. Although the thirteen- or fourteen-year-old girl may be emotionally and psychologically a child, her body, based on its fat content and weight, has decreed her to be a woman, capable of child-bearing. Puberty is oblivious to her social state.

Animals other than humans must bow to the demands of pregnancy as well, and, if underweight or poorly nourished, they suffer delays. Even though their reproductive equipment may be ready to roll, rabbits, red kangaroos, heifers, dogs, and even birds like chickens, all hold pubescence at bay until the proper weight.

Furthermore, just as poor nutrition slows weight gain and can delay puberty, enhanced nutrition can speed up its onset. Plenty of acorns do not always grow a forest of oaks but may make a squirrel pubesce at four months instead of the usual ten months. Red deer, transplanted from their native barren highlands of Scotland to the lush evergreen forests of New Zealand, speed up puberty from three years to sixteen months. One fawn, who was fed a delectable, highly nutritious, high-protein diet of alfalfa and clover, sprouted a mature follicle at four months and tottered on the brink of infantile pubescence. Similarly, in other artiodactyls, such as moose, mule deer, and wapiti elk, puberty onset is enhanced by an abundance of high-quality vegetation. In years of abundance, young females pubesce earlier because they gain weight sooner, thereby increasing the population faster.

As in the pubescent girl, fat is beautiful, even among wapiti elk or deer. Without fat, there is no motherhood.

But for many other species, the season of the year regulates puberty. The first pregnancy (and all subsequent pregnancies) must be synchronized with the seasons so that the offspring will be born into a world filled with lush vegetation, not into one that is starkly bare, harsh, and wintery. Many small rodents breed for the first time during their first spring.

One such creature is the field vole. Her entrance into adulthood does not depend on her age or her weight, but rather on the lengthening days that anticipate the summer and its burgeoning food supply soon to be available to her young.

Among female sheep, the time of puberty onset bows to the requirements of posterity, even causing early pubescence, if necessary. Young ewes enter puberty together during the shortening days of fall, breeding then for the first time and lambing together five months later in the spring, again a propitious time for growth. Then, too, the synchronized birth of lambs results in a greater chance for their survival, since all the ewes are mothers at the same time and can cluster together, each one alert to potential predators, warning the others of imminent danger.

The entrance of female fin whales into whale adulthood is controlled by lengthening days but is somewhat complicated by the whale life-style. Fin whales migrate longitudinally, from the Arctic to the Antarctic and back, crossing the equator twice a year in such a way that they always enjoy the summertime, cleverly avoiding wintery climates. North or south, the direction of migration is immaterial, because the females pubesce when the days grow long; hence some pubesce on their northerly route and some on their southerly migration. Young are born into a perennial summertime.

Many bird species, for whom lengthening days are the mentors of sexual maturity, also follow the strict dictates of the season, producing young in time to garner the seasonal abundance of spring.

Young Love and Perfect Timing

Timing puberty to the right season is evolutionarily smart, as is timing it to the right diet and weight. But one's social interactions may also be critical. Whom one lives with and how many roommates one has may tamper with one's sexual maturity. For example, for mice that are packed together into laboratory ghettos with hardly any breathing space, puberty is delayed. Tiny field voles, whose population densities soar and crash every three to four years, delay maturation when they are overcrowded, like the laboratory mice, but reach reproductive age earlier when the population is scarce. Such small mammals are not the only ones that suffer from overcrowding. Dense populations of African elephants in the Murchison Falls Park of Uganda reach puberty at about eighteen, while in other herds with fewer members, females reach sexual maturity at eleven or twelve. Unfortunately, crowding and poor nutrition often go hand in hand, and the observers have left us dangling, for we do not know if the overcrowded elephants or voles had enough to eat. However, laboratory evidence suggests that overcrowding may take precedence over nutrition as a delayer of puberty among wild animals, just as it does in laboratory mice. Although well-fed, crowded female mice hesitate to enter adulthood.

In the natural world, crowding is an unhealthy condition, meaning that there are too many mice for the amount of available food. Producing young in this impoverished environment is a poor reproductive strategy, since starvation looms in the offspring's future. Evolution has made female mice "smart"—capable of dampening their reproductive potential, at least for a while. Females secrete a "crowding" chemical in their urine, which serves to mutually inhibit the onset of sexual maturity.

In laboratory experiments, supervised by John Vandenbergh, female mice were raised under a variety of social conditions to see what other factors affected the onset of sexual maturity. When reared with males, females came into puberty sooner than isolated females and much sooner than crowded ones. Vandenbergh also found that the urine of either sex contains pheromones, which by acting in a way as yet unknown influence the age of puberty. Other rodents seem to show similar patterns but this may not be a widespread phenomenon. Would a physiologically ready girl who has five older brothers pubesce earlier than one with five older sisters? No one has studied this issue yet.

Although the actual physiological and anatomical changes of puberty are programmed in the genes, the timing of puberty is under the sway of the environment. This adaptation allows the female to speed up or delay her first pregnancy so that she can make the best use of all available resources: extra food, warm weather, a sexually inclined male.

Among most mammals, the first sexual heat brings with it small anatomical and huge behavioral changes, along with the beginnings of that special female trait, the cycling of sex hormones. Permanent anatomical changes are minimal and barely discernible. Who would notice that the pubescent rat has opened her vaginal membrane, readying it to receive a penis, or that the clitoris in the spider monkey enlarges, or that the pelvis in the squirrel monkey widens?

Puberty is signaled more overtly by striking new behaviors and the emergence of a vast array of new broadcast scents, expressing the dramatic underlying hormonal changes and, with them, the desire of the female to have a sexual encounter and copulate.

For example, the way we know our kitten has come to the end of her kittenhood is by the way she behaves when she comes into heat for the first time — by her excessive craving of bodily contact and her plaintive vocalization. Pet her back and she arches it, crouching forward to reveal her hindquarters and expose her vagina with its seductive feline musk. A pubescing female gerbil suddenly darts past the male with stiffened legs and her tail straight out behind her. Fluffing her fur to make herself conspicuous — and perhaps even attractive — she reminds us of women teasing their hair. The male follows the female and presses his nose into her vagina. Turned on by the male's interest in her pungent vaginal aromas, she crouches with her rear thrust in his face and awaits his mounting.

Many behaviors shown by a female at puberty and in subsequent heats are designed explicitly to spread around her sexual perfumes and thus increase her chances of attracting a male willing to copulate. Perhaps one of the least subtle of all mammals is the New Forest mare studied by Stephanie Tyler in the Sub-Department of Animal Behavior at Cambridge University. After many months observing social behavior, she reported differences among older females as compared with the younger, pubescent females. In her maiden estrus, a pony mare stands with her tail arched upward, hind legs spread slightly apart, and protrudes her clitoris repeatedly through vulval lips, a behavior aptly called "winking." If a nearby stallion rushes excitedly to her, she prances in front of him, spurting urine, which carries a horsey perfume, and swishes her tail from side to side to spread the scent around.

A pubescent female rhesus monkey emits a sweaty-socks odor from her vagina that, although repulsive to us, seduces the male. She does not stop at smell alone, but appeals to the primate's sense of color by making her bottom an eye-catching, blushing red, hardly to be missed, even by a male whose sense of smell is diminished by a raging head cold.

Young female lowland gorillas may slightly swell their anogenital regions, but they also announce their puberty by some hard-to-miss signals, such as chest beating. As Gary Mitchell reports in *Behavioral Sex Differences in Non-Human Primates*, "they 'mouth-groom' young males by kissing them and sticking their tongues in the males' mouths ... The pubescent or sub-adult female gorilla also rides the back of a male for anywhere from 5 seconds to 10 minutes during which time the male has a visible erection." Such pubescent females initiate almost all seductive foreplay.

But pubescence in the female human, as we mentioned, is indicated in a way completely different from all other mammals, with permanent and striking anatomical changes; the swelling breasts, underarm and pubic hair are all unmistakable markers. But unlike her mammalian cousins, she does not signal puberty with an exhibitionistic first sexual heat. The onset of adulthood is marked not by behavioral changes but physiological ones, when she begins to menstruate.

The Menstrual Cycle: Harmony of Body and Brain

On the first day she menstruates, a girl is considered to be a woman with reproductive potential. That is her social puberty. With menstruation comes the awakening of the eggs that have lain dormant in the ovary since before birth. The first menstruation, like the first sexual heat, marks the onset of hormonal cycling.

Sex hormones previously produced in small, constant amounts now fluctuate with great swings, rising and falling cyclically.

Female hormonal cycling directly affects three parts of the body —the ovaries, the uterus, and a small part of the brain called the hypothalamic-pituitary center. This female trinity, although physically spread out between pelvis and head, is actually in constant contact through the chemical conduit of sex hormones. Often misunderstood, intrusive, noticed mainly as an inconvenience, the cycle is really so simple that to understand it may lead to loving it, or at least disliking it less.

The events of the cycle are chemically controlled by the fluctuating production of hormones in two parts of the female trinity: the hypothalamic-pituitary complex, in the brain, and the ovaries. Together, the brain hormones and the ovarian hormones control all the steps of the cycle, each leading naturally to the next, forming a unit that completes itself in about twenty-eight days. Within the completion of one cycle is the beginning of the next. Although menstruation is actually the last phase of any cycle, it is a convenient signpost, because it is the one event that makes itself obvious—through its profuse, often uncomfortable, bleeding.

The main ovarian hormones, called estrogen and progesterone, are composed of tiny, compact steroid molecules derived from cholesterol, that infamous bugaboo of the health-conscious. Brain hormones—larger, more complex proteins—are gonadotrophins, so named because of their affinity for gonads.

Few other substances in the body are measured in units as small as the units used to measure hormones—nanograms and picograms. A nanogram is one-billionth of a gram, and a picogram is one-trillionth of a gram, an unimaginably minute amount when you consider that there are 454 grams in a pound. To emphasize how few sex-hormone molecules are in your body, we would need the blood of five thousand women to obtain one teaspoon of progesterone; to obtain one teaspoon of estrogen, we would need to drain the blood from over two hundred thousand women.

Hormonal potency is without peer. It is rather terrifying to think that the future of humanity rests upon a few molecules of hormones released from the hypothalamic-pituitary complex and from the ovary, a couple of glands no bigger than a big toe. It overwhelms one to think that so few molecules control such major events and that a few molecules more or less can wreak physiological havoc. Even more remarkable than hormonal potency is the close communication between brain and ovary that oversees the cycle.

An example of what would happen if we did not have hormonal concordance and modulation between brain and ovary comes from an intensively studied tiny parasitic wasp, *Habrobracon*, whose genes alone control the expression of its reproductive physiology and behavior. In normal mating, the female seems to be barely aroused, passively accepting and copulating with the ardent courting male. Although showing no interest in the attentive male, this two-millimeter-long wasp becomes aroused, indeed inflamed, if after copulation she sees a caterpillar of the cornmeal moth. Pursuing the caterpillar, she stings it with toxins and paralyzes it. She lays her eggs on it, and it serves to nourish and protect the eggs. Her larvae eat their way through the paralyzed caterpillar as they grow into parasitic wasps.

This wasp is a splendid experimental animal to prove how essential the brain-ovary hormonal communication is. The wasp's chromosomes can be manipulated to produce the brain of one sex, the body of another — hence an animal of ambivalent gender. A female-brained wasp, heedless of its sperm-filled testis, mates normally, passively accepting a courting male and, after mating passionately, pursuing a nurturant cornmeal-moth caterpillar, which it stings and prepares for its nonexistent eggs. In reverse, a male-brained wasp raises its macho sexual flags but, although egg-laden, ignores the nurturant caterpillar. The wasp can be genetically manipulated even further, for its brain is highly susceptible to chromosomal and genetic manipulation, leading to some real freaks, truly confused wasps who stupidly sting their own females instead of the caterpillar and foolishly try to copulate with the caterpillar.

These aberrations arise because the brain and gonads are not acting in concert, are not "talking" to each other to coordinate physiology and behavior. Genetic and hormonal information are treated as two separate entities; hence, we can create an egg-filled creature who acts like a courting male.

To have hormonal concordance and modulation requires time. Since hormones, compared to nerves, are slow, rather plodding messengers, time is the essence of cycling. It takes time to mature an egg, and it takes time to prepare the endometrium and, if necessary, to destroy it. For a typical cycle, twenty-eight days elapse.

Twenty-eight days — the echo of the lunar month, encompassing the phases of the moon; the echo of the ebb and flow of the tides. Even the name of the cycle echoes the Latin word for month: *menstruus*. Surely one can be right in assuming that the approximation of the twenty-eight-day cycle to the phases of the moon has contributed to the supernatural aura and to the "lunacy" with which menstruation is greeted in many cultures. Spontaneous bleeding is generally an unhealthy sign, and the superstitious might believe that its regular recurrence every month must be the work of the devil. In fact, many Western women still call this natural and essential phenomenon "the curse"—a sentiment not so far removed from the attitudes of more primitive cultures who exiled menstruating women, declared them unclean, untouchable, and made them the object of social abuse. In *Sex and Temperament in Three Primitive Societies*, the famed anthropologist Margaret Mead, documenting the social condition of menstruating women in many cultures, reported that Arapesh women of New Guinea are banished to a small poorly constructed shelter on the edge of a hillside and that their husbands must take over full responsibility for the family. Among the Manus women of the Admiralty Islands, menstruation is regarded as "so shameful that it must be hidden," and their word for this biological event means "leg-broken." Among the Balinese, a menstruating woman is considered contaminated and is barred from religious worship.

But the same cultures that malign menstruation also exalt its onset, the symbol of awakening fertility in the girl.

For example, a pubescent Manus girl, according to Mead, celebrates her first menarche, or the onset of menstruation, by "a great ceremony; the other girls of the village come to sleep in her house, there are large exchanges of food and ceremonial and splashing parties in the lagoon; men are excluded and the women have a few jolly parties together—then absolute secrecy descends on the girl's later menstruation." Thus, the cycle's onset is worshiped, and subsequent ones, strangely enough, are abhorred.

Mead also wrote extensively about the menarchal ceremonies of pubescent Arapesh girls, who enter womanhood through lengthy ceremonies lasting for as long as a week. The Western girl has her own simple ritual. She may invite a few friends with her on her trip to the supermarket to buy her first box of sanitary napkins, giggling all the way.

In current Western culture, many people abstain from intercourse during menstruation because of its messiness and sometimes as a result of religious beliefs. In fact, the Old Testament clearly states that a menstruating woman is unclean and impure.

Whether it is the bleeding itself or the periodic nature of menstruation that has contributed to the bizarre beliefs and superstitions surrounding it is unknown — undoubtedly both are culprits. Yet menstruation is neither strange nor bizarre; it is part of a cycle that is of fundamental importance to the future of our species.

There's nothing mystical about the month-long cycle, about the two weeks of preovulatory preparation and two weeks of postovulatory expectations, and there is nothing mystical about physiological systems and the ways in which the body's parts communicate with one another.

The exchange of chemical correspondence between brain and ovary does proceed rapidly because each organ is constantly "checking out" the hormonal condition of the other one. Within the cycle, there are many steps, each requiring the appropriate time for its completion (e.g., maturing an egg, triggering the egg's release, synchronously enriching the endometrium, rolling an egg down one of the fallopian tubes, and shedding the lining if the egg is not fertilized).

Twenty-eight Days — a Close-up

Let's go through a hormonal cycle, starting at an arbitrary point — arbitrary because the cycles by nature are continuous. Conventionally, the cycles of a female human are measured from her first day of menstruation. During the nuisance-filled first few days of bleeding, which indicates the shedding of the now-useless endometrium, estrogen, progesterone, and the gonadotrophins are at their scarcest. The hypothalamic-pituitary center, sensing very few estrogen molecules in the blood, now sets about to correct their scarcity by producing a special gonadotrophin called FSH (follicle-stimulating hormone). Pouring into the blood, FSH molecules travel to the ovaries where they prod follicle cells to ready the dormant egg for ovulation. The effervescent follicle cells not only prepare the egg for release but also secrete their own hormone, an estrogen, which pours into the blood as well. With this outpouring, the number of estrogen molecules circulating in the blood increases, and the increased level tells the brain that the body is ready for the next phase in the cycle. The brain then reduces the manufacture of FSH and kicks in a new gonadotrophic hormone, LH (luteinizing hormone).

The sudden deluge of LH springs open the follicle, casting the fertilizable egg out into the fallopian tube. The brain then halts its output of LH, and the follicle cells collapse inwardly into the space formerly occupied by the egg. The follicle transforms itself into a yellowish blob of cells, called by the unimaginative name of corpus luteum, "yellow body." This body replaces the follicle as the major maker of ovarian hormones and adds to its hormonal retinue a steroid called progesterone, so named because of its progestational function.

Together progesterone and estrogen prompt the third member of the female trinity, the uterus, to grow a spongy, blood-rich endometrium. The endometrium, which has a short life span, is in its peak condition a few days after ovulation, not coincidentally the time when the embryo will arrive at this rich, well-stocked tissue if the egg is fertilized.

If an embryo implants in the warm, inviting uterine haven, then progesterone continues in its progestational role, seeing to it that the uterus and fetus grow placental connections to keep the nutrients flowing and the wastes removed throughout the nine months of gestation.

If no such embryo arrives, the uterus disencumbers itself of the now-useless nutrient-filled lining; this process is menstruation. Progesterone and estrogen secretions subside, and the woman experiences her monthly bleeding.

Normal cycle length commonly varies from woman to woman and may range from twenty-one to about forty days. This variety merely reflects differences in the preovulatory phase: Some women take scarcely a week to build up a uterine lining, while others take three. In the same way, deviations or irregularities in the cycle lengths of an individual woman also reflect variations in the length of the preovulatory phase. This part of the cycle is extremely sensitive to outside influences, as well as internal physiology. Stresses such as episodic emotional disturbances, minor illnesses, final exams, jet lag, social events, and extreme weight loss or gain can dramatically alter the length of the first part of the cycle.

In contrast to the sensitivity of the preovulatory phase to potential disturbances, the postovulatory phase is much more uniform, almost always taking two weeks. It is relatively unaffected by emotions —less influenced by fights with one's husband, a trip to Africa, or the common cold.

The female human could be a role model for reproductive cycling in other mammals, for the hormonal events and the internal changes are the same, but the female human differs significantly in the way the cycle's existence is recognized.

Some of our primate relatives also experience periodic bleeding, and if they had the capacity, they might be able to mark their menses on monthly calendars. The rhesus monkey, for example, upon reaching monkey womanhood, menstruates every twenty-eight days, just like us. However, although monkeys menstruate, it is an insignificant event, overshadowed by a more important event taking place at a different time during the cycle — the periodic sexual heat, which happens at the time of ovulation.

Although many anatomical and physiological similarities connect humans to their evolutionary relatives, Old World monkeys, the display of sexual heat is not a shared characteristic. Most of our primate relatives are more like other mammals, who signify pubescence and cycling onset with their first estrus. With pubescence begins the remarkable three-way hormonal communication among the trinity of brain, ovary, and uterus.

The physiological events of an estrous cycle are mirrored in the menstrual cycle. Basically the only difference is the difference in markers: in essence, they are the same cycle. Both cycles are ovulatory cycles, synchronizing with egg release the peripheral support systems needed to nurture a developing fetus.

Pubescence, the biological end of childhood, brings with it the burdens of adulthood and the pressure to fulfill the reproductive mandate, carved into the brain and body so very early in life.

4
Timing Is Everything

A HUMAN BABE takes nine months to gestate, and consequently, those of us who plan ahead try to conceive so that the baby is born at a convenient time for us, whether it be one day of a hot summer, the end of the spring semester, or the middle of a frigid winter. The earth's seasonal productivity rarely enters into our plans of birth time, for markets are stocked with foodstuffs all year long — easily available if one has money. We are the only species with such freedom. All other species — about two million of them — must rely directly on the earth's seasonal bounty for nurturance of their offspring.

Newly hatched, newly born, newly metamorphosed, each and every offspring needs to eat. That simple patent fact is singularly instrumental in the evolution of a vast array of female "time and place" strategies. Not only must females produce eggs, but they must produce them at the right time and in the right place, for this will have a crucial effect on the egg's future. In the interest of species survival, the anticipated availability of food is just as important as getting the egg fertilized.

Female adulthood brings with it a heavy burden. Species reproduction is her biological bottom line. As an adult, she must stretch her concerns far beyond her narcissistic need for food and safety. She must attend to myriad stimuli, heretofore irrelevant, and she must fit her own body even more carefully into the complex physical and social environment that makes up her habitat.

But the world is often a difficult, unfriendly place. To compensate for the stringent demands of a hostile habitat, evolution invents all sorts of clever strategies so that such species can survive and reproduce amid the rigors and stresses of the environment. Thus, in addition to concentrating on the many details of day-to-day survival, the demands of reproduction force the female to become attuned to new outside cues and to the changes occurring inside her body as well.

By "knowing" the time she must give birth, the female can prepare her body and her eggs sufficiently in advance so that the production of young will proceed according to species schedule. Whether she comprehends her awareness of new, pertinent environmental information, as we do, or whether her genes are "programmed" is irrelevant to our discussion. Evolution has endowed females of all species with the capability to sift through and to select from many different stimuli those that are critical to forecasting the right moment for hatching or birth. The emergence of young from membrane, shell, ovary, or uterus determines how and when the strategies of reproduction begin.

Time can be friend or foe. The truism "Timing is everything" is bandied about all the time, referring to good or bad luck, to being in the right or wrong place at a certain time. But while that notion may seem trivial to us, among animals timing can be truly a matter of species life or death, even in the first steps of reproduction. If mature eggs are not produced at propitious times, species death looms menacingly. Indeed, perhaps such species as the dinosaurs became extinct because their eggs and subsequent offspring arrived on the scene out of sync with a beneficent environment.

Synchrony and harmony of the animal with its particular environment — that is, its niche — is a part of nature over which we marvel, usually without thinking of the many steps, the many confluences that went into that magnificent integration. With casual assurance, we agree that the right time for egg release and the mechanisms that bring it about are part of the species' nature. But so what? Such a statement tells us nothing about the finely honed systems, about the fact that females must go to great effort, expend huge amounts of energy, even put themselves in danger, to harmonize their bodies with their environment's demands and to derive from it every conceivable benefit for the well-being of their offspring.

The habitat of female vertebrates bombards them with information about season, changing day length, temperature, phase of the moon (among other variables), as well as bombarding her with information about the chemistry and behavior of potential species mates. Overwhelming as all this information seems to be, the female is effectively programmed to prepare eggs in advance, release them at the right moment, and, where relevant, hatch or birth her young propitiously.

For example, a female bird is genetically programmed to lay eggs. She has all the needed physiological accoutrements to manufacture them, enclose them in a shell, and pass them into a nest from her body. She is even programmed to respond to signals from the male with postures, vocalizations, and other behavioral signals of her own. Outwardly unremarkable, the steps in the process of mating and egg laying can appear to proceed in a scheduled sequence. Some observers have compared such behavioral sequences to the flushing of a toilet. Once the cord is pulled, the rest of the sequence spills forth, untouched by other stimuli, by attempts to stop it, change its speed, or alter its eddy. That simplistic metaphor explained everything so no further comprehension of the reproductive behavior seemed necessary. These researchers were satisfied that internal mechanisms were the guiding forces of a species' behavior.

The Ring Dove: Feedback Systems at Work

They ignored the elegance of an evolutionary work of art: The feedback system which allowed the female to time her egg release and to alter her physiology. It was ignored until a psychologist at Rutgers University, Daniel Lehrman, and his colleagues, scrutinized during the 1950s and 60s the mating behavior and ovulation in the ring dove, a pretty pink-beige bird. He showed that the sequences of mating and egg laying did not "pour forth" like the proverbial water spilling over a dam but were closely attuned to the quality of the external stimulus. He found that the female even altered her own hormone levels, essential to continuing the sequence.

Lehrman's observations of the typical behavior of mated birds led eventually to a series of experiments heralding a scientific breakthrough and restating what philosophers have so often said of humans: We change our environment, and our environment changes us.

According to Lehrman, in an article in *Scientific American* in 1964, "If we place a male and a female ring dove with previous breeding experience in a cage containing an empty glass bowl and a supply of nesting material, the birds invariably enter on their normal behavioral cycle, which follows a predictable course and a fairly regular time schedule. During the first day the principal activity is courtship: the male struts around, bowing and cooing at the female. After several hours the birds announce their selection of a nest site (which in nature would be a concave place and in our cages is the glass bowl) by crouching in it and uttering a distinctive coo. Both birds participate in building the nest, the male usually gathering material and carrying it to the female, who stands in the bowl and constructs the nest. After a week or more of nest-building, in the course of which the birds copulate, the female becomes noticeably more attached to the nest and difficult to dislodge; if one attempts to lift her off the nest, she may grasp it with her claws and take it along. This behavior usually indicates that the female is about to lay her eggs. Between seven and eleven days after the beginning of the courtship she produces her first egg, usually at about five o'clock in the afternoon. The female dove sits on the egg and then lays a second one, usually at about nine o'clock in the morning two days later. Sometime that day the male takes a turn sitting; thereafter the two birds alternate, the male sitting for about six hours in the middle of each day, the female for the remaining 18 hours a day.

"In about 14 days the eggs hatch and the parents begin to feed their young 'crop-milk,' a liquid secreted at this stage of the cycle by the lining of the adult dove's crop, a pouch in the bird's gullet. When they are 10 or 12 days old, the squabs leave the nest, but they continue to beg for and to receive food from the parents. This continues until the squabs are about two weeks old, when the parents become less and less willing to feed them as the young birds gradually develop the ability to peck for grain on the floor of the cage. When the young are about 15 to 25 days old, the adult male begins once again to bow and coo; nest-building is resumed, a new clutch of eggs is laid and the cycle is repeated. The entire cycle lasts about six or seven weeks and — at least in our laboratory, where it is always spring because of controlled light and temperature conditions — it can continue throughout the year."

Lehrman and his coworkers concluded that ovulation in this species results only after the completion of a series of external events and that genetically generated processes were not enough to direct the sequence. The female requires these environmental resources, in order:

the day length of spring; the sight and sound of a bowing, cooing male; materials to build a nest; an additional week or so to produce eggs. Her hormones, along with her behavior, responded to these stimuli completing the scenario for reproduction.

To better understand female behavior, Lehrman and his associates divided the behavior into sections to determine which parts were critical. They discovered that her brain didn't start to produce the hormones of reproduction until she accepted a courting (bowing and cooing) male as her mate. The accepted male stimulated her hormone production, which then triggered a new behavior, nest building; she decorously arranged the straw in a glass bowl, conveniently provided by the experimenter. The act of building the nest — the action of picking up each straw and arranging it just so until the nest was completed — was the stimulus that told her brain it was time to make the hormone of ovulation. She mated, laid her fertilized eggs, and incubated them in the nest.

This was a remarkable discovery: Nest building behavior is not just the slave of this bird's physiology, to be switched on when hormones so decree, but is the master of physiology as well.

The significance of Lehrman's research was to make other scientists aware that the female ring dove does not lay eggs by the simple triggering of a genetically programmed timer. She does not start to prepare eggs or even ovulate until all the correct conditions (the right sound of a bowing; cooing male; a nest) are met. If they were not met, ovulation would be pointless and she would have expended enormous amounts of energy without producing any viable offspring.

Subsequently, Robert Hinde, an animal behaviorist at Cambridge University in England, found that canary behavior and physiology echoed that of the ring dove; a female held her eggs in abeyance until she heard the lilting songs of a courting male. Then, "knowing" she had a male nearby who could fertilize her eggs, she built and literally feathered her nest.

These birds only serve to exemplify the behaviors of other species, whose females all time ovulation to a specific situation, retaining their eggs within their bodies until they receive appropriate release signals from the physical and social environment, as well as from internal stimuli.

Eggs Welcome Here

The black-chinned mouthbrooder, a favorite fish in the laboratories of the American Museum of Natural History, engages in an elaborate courtship during which she and her consort pass one another with amorous intent, lock jaws in a piscine kiss, and energetically scoop out a nest in a gravel bed. Their amours last for several days, giving the female time to mature her eggs. When the hollowed-out nest is finished, the sight of it triggers the next phase of her reproductive behavior: She and her consort alternately swim over the nest, touching their genital tubercles to it. Satisfied that the cleaned nest is the right shape and size and that an amorous male with plenty of sperm is attending it with her, she ovulates.

Despite the days of preparation and the energetic scooping out of its gravel, the nest of this species serves the female only briefly, for as soon as the male fertilizes the eggs in it, he sucks them into his mouth, where they spend the next ten days growing into little fish.

Among these laboratory fish, the earth's bounty was irrelevant. Fed every day by an attentive aquarium keeper, they spawned in warm greenhouses during the entire year, while their wild cohorts in the lagoons of Nigeria spawned only during the African summer.

Amphibians must also time their egg laying to appropriate environmental stimuli; in the case of the California newt, rain water is the cue. These females live their lives under fallen logs on the forest floor, but their larvae must hatch in a stream or pond. While still leading a solitary life under the redwoods, the female begins to respond physiologically to the shortening day length and cooling temperatures of autumn by making hundreds of large yolky eggs.

When the first heavy downpour of the winter rainy season comes, she and hundreds of her kin drag their swollen bodies down to a stream, singlemindedly pursuing their ultimate goal of successful reproduction. In the stream, she is courted by males, who crawl over her body and caress her during their underwater orgies. Stimulated by the love play of courtship, her cloacal lips pick up packets of sperm left behind by the salamander swains. Then she ovulates, and as the eggs leave her body through the cloaca, they pass over the fresh sperm and are fertilized.

The Maine lobster cues time of egg release with time of mating.

Normally encased in a hard chitinous shell, as she grows she periodically molts, that is, casts off her shell, increases in size, and makes a new hard shell again. But while she is in the soft-shell stage, she mates. Molting can take place without mating, but mating cannot take place without molting. A solitary type, she is in considerable danger of becoming prey for another animal when she leaves the safety of her burrow. Nonetheless, despite the ever-present threat of death, she departs from her rocky burrow and seeks out a reclusive male in another burrow. She must persuade him, by appropriate courtship signals, that she does not want to inhabit his burrow or want to eat him. Convinced of her romantic interest, the male mates with her by inserting special copulatory organs, which transfer a packet of sperm into her genital aperture. Then, as she hardens her soft shell, she encases and stores the sperm packet in a special receptacle within the new armor.

After she sheds her eggs, the female keeps them tucked against her ventral surface, holding them secure with her appendages — in effect, creating a portable nest.

In sum, many vertebrates closely coordinate time of egg release with a specific sequence of signals, such as the presence of a courting male and the building of a nest. If no rains came, the newt would not begin the series of events leading to reproduction. Without a nest and an attending male, the African mouthbrooder would not lay eggs. Without the proper stimuli in sequence, birds, alligators, turtles, and numerous other species would not release eggs even after making them, and some would not even make the eggs.

Evolution, by placing the protective, nurturant nest within the body of the female, allowed mammals to make a significant evolutionary leap, adapting to a huge variety of habitats — swimming with fishes, flying with birds, as well as inhabiting the traditional niches of terrestrial vertebrates. For example, the demands made on a polar bear hunting on the glaciers of Siberia are obviously very different from the problems faced by a dik-dik browsing through the acacia plants on the African savannah, and the demands from both of those niches are vastly different from the stringencies faced by the sea otter searching for delectable abalone in a kelp forest off the Monterey Peninsula.

Yet all their embryos' worlds are alike. Consistency is bestowed by the cyclical events that make the uterus a protective nurturant nest, readied at the right moment to receive a fertilized egg. Other than that common bond, the female of each species fits her life-style to obey the stern taskmaster of habitat and, consistent with other vertebrate patterns, birthing time obeys the cyclical nature of the earth's bounty, and mating time anticipates gestation periods.

Thus, if the species gives birth only in April or May and gestation is six months long, obviously mating should occur in October or November. Such females follow the mandates of a breeding season, not cycling—ergo, not producing eggs—at any other time. An error in timing could be energetically very costly to her, and her offspring would have little or no chance of survival if, for example, it were born in January into a frigid winter storm.

Like birds and other vertebrates, mammals are tuned in to such environmental stimuli as day length, changes that are communicated through the sensory pathways and into the brain to create internal changes, leading to onset of the ovulatory cycle. Yet despite her freedom from fabricating an external nest, the female mammal also is constrained.

Females that build external nests apparently react to the sight of the finished nest and then shed their eggs. The mammal also sheds her egg when her internal uterine nest is fully prepared, but she cannot see her nest and has little or no overt control over its completion and, in most species, cannot control when the egg is released. Its release is timed to happen when the nest is finished. But she is still faced with the same problem as her vertebrate cohorts. She must put herself into the right place to get her eggs fertilized, and she must get them fertilized at the right time during the breeding season. She resorts to another strategy to announce the presence of a ripe egg and, coincidentally, "nest" readiness.

The Red Deer Harem

The way a female behaves to put herself in the right place at the right time is beautifully demonstrated by the classic study carried out by F. Fraser Darling, an Associate at Edinburgh University (Scotland), who spent two years during the 1930s following shy, secretive, wary

TIMING IS EVERYTHING

red deer as the herd migrated up and down the hills of Scotland eating the abundant seasonal vegetation. The right time for mating is fall; the time for birth of young in the spring; the right place within a stag's territory, in the hills.

The breeding season, often referred to as the rutting season (clearly a man's term), starts at the end of September. The females are already ensconced in the male's harem, eating, sleeping, and milling about together under his watchful territorial eye. Then the females begin the first of their eighteen-day estrous cycles, going full-blast into their preovulatory phase. As the egg is readied for ovulation, the female enters the state of estrus, and since this state lasts but one day, she must announce it, which she does in clear and wonderful ways. She teases and caresses the male, indicating her willingness to mate. In *A Herd of Red Deer*, Fraser Darling describes one hind's behavior as follows: "The stag approached her; she ran away; he chased her; she stopped and came to him; rubbed her whole length along his ribs from fore to hind end; made as if to mount him; he turned to mount her; she ran away; he chased; she stopped; he mounted and served her. When he was sliding off she kicked up her heels, hit him in the belly, and ran away. He roared and followed her again; she came to him and rubbed herself along each side; licked his muzzle, walked under his chin, throwing back her head, licked his sheath for a moment, made as if to mount him; he turned, mounted, and served her again."

If, after all these antics, the male does not fertilize her eggs during her one day of heat, she completes the cycle and enters another one soon afterward, becoming sexually interested in the male eighteen days later. Her estrous cycles may continue past the peak rutting season, as long as six months or ten cycles, until she is impregnated or the breeding season terminates. However, it is a rare hind who is not started on motherhood during the first month of breeding when deer passion is at a peak.

Female red deer get together with males only during the breeding season. Choosing to remain in a stag's harem assures them of having a stag handy when their sexual heat peaks and the eggs are ready to be fertilized. At the end of the breeding season, when their cycling also comes to an end, they move off en masse, abandoning the stag—who is no longer of any use to them.

By not cycling during the other part of the year and therefore not having any eggs, the female guarantees that her offspring will be born during the next spring, when food resources are abundant and weather is favorable.

The red deer has a rather long breeding season, but, as mentioned, other female mammals must be impregnated during their one and only estrous cycle or wait an entire year for the cycle to recur.

A year is a long time in the life of a mammal, so the female employs strategies that will assure her of having a male available at the time of her one and only estrus that year. A red fox, for example, grudgingly permits a male to follow her, traveling the same routes and sleeping near her for one or two months before estrus. Normally, the vixen is a solitary creature, hunting and roaming the woods by herself or with her daughters, eschewing the company of males. But she turns her other cheek as part of her reproductive strategy and even engages in a love affair replete with vulpine kisses and ear nibbles. Mating evidently takes place during the one night of her estrus, and, after a foxy one-night stand, the male leaves the following morning.

It may seem strange to us that the vixen mates during the shortest days in the snows of January and the red-deer hind during the shortening days of fall. We think of spring as the time when love blooms, when a young man's fancy lightly turns to thoughts of love, when the insects chirp and sing their arthropod lovesongs and bright yellow birds warble as they build nests on treetops. Yet spring is not the time of mating for all animals. For mammals with long gestations, it is the time of abundance, of flowers, of grasses, of insects — and of baby animals.

The estrous cycle of these females is classic, usually eighteen days long, following the standard pattern of three phases: preovulatory, ovulatory, and postovulatory. Each phase follows its fixed course, running itself out according to the traditional physiological system. But for some species, it behooves the females to make alterations upon the standard pattern, to make adjustments to suit her life-style, nipping and tucking, shortening one phase, jettisoning another, keeping eggs unreleased until a male is around.

One animal with a very brief estrous cycle indeed is the laboratory rat, probably abbreviating it because, like other small mammals, she doesn't live very long. She manages to compress the cycle to five days from its typical eighteen-day length by casting out the postovulatory phase. She does not manufacture a corpus luteum unless she mates. If unmated, she starts the preovulatory phase of the next cycle immediately, producing fresh eggs five days later. Oddly enough, if the female has mated and the eggs are infertile, she responds as if pregnant, makes a corpus luteum, and hence has a postovulatory phase, even building an external nest for her anticipated young who will never arrive. Many other rodents, tiny short-lived mammals that they are, are forced to reproduce rapidly and can't afford to wait three weeks for the next batch of eggs; they, too, eliminate the last phase. (Actually, it's energetically very efficient not to bother to build a uterine lining unless an embryo is around to make use of it.)

Nature might have been smarter to give these small mammals an even more efficient system, however. Instead of ovulating even if there is no male around, they might have evolved the strategy of certain other species, who ovulate only when there is a male present — the most efficient use of time.

Eggs On Demand

The champion mammalian reproducer, the only one that breeds like a rabbit, is the rabbit. The rabbit and a few mammals we'll mention later have modified their estrous cycle by inventing an eggs-on-hold system. The female always has an egg ready to release and ovulates only when she mates. Although she has a typical preovulatory phase during which a follicle matures, the follicle cells retain the ovum until they receive the pertinent message; mating has brought sperm into the vicinity of the egg.

Rabbits continuously mature eggs and can ovulate at any time, in contrast to the traditional ovulators, whose timing is restricted. Yet rabbits have not overcome the perishable nature of eggs, and as some deteriorate from egg "old age" within their follicle, she simply replaces them with fresh eggs.

The rabbit is constantly in heat and thus produces an abundant share of bunnies. The often perjorative expression "breeding like rabbits" reflects the doe's sexual appetites and her willingness to start making baby rabbits whenever she stumbles across a male. Her fresh egg supply once made her indispensable in early pregnancy-detection tests. Medicine took advantage of her eggs-on-demand reproductive strategy by developing the famous rabbit test: A little urine from a possibly pregnant patient is injected into a female rabbit. If the woman is pregnant, her urine contains a hormone which causes ovulation and the formation of a corpus luteum in the rabbit. The euphemism for pregnancy—"The rabbit died"—meant that in the dissected ovary of a sacrificial doe a corpus luteum had been found. Pregnancy was confirmed. (However, rabbits were expensive to rear and maintain, so they were eventually supplanted for pregnancy tests by another egg-on-demand animal, an induced ovulator, the leopard frog. It spewed its eggs out into the water; did not need to be dissected; was very inexpensive to feed; and had no endearing qualities, so its passing was never mourned. Now leopard-frog and rabbit tests are a thing of the past, replaced by a simple tube of antibodies that coagulate into a telltale ring, revealing pregnancy.)

The female rabbit lives with lots of males, but other practitioners of the egg-on-demand strategy are solitary—hunters such as cats, ferrets, minks, or homebodies such as moles, who rarely leave their single-dwelling burrows. These solitary females also modify their estrous cycles, holding back egg release until the rare time that a fertile male is around.

Strange as it seems, camels are like ferrets. Although camels are now domesticated, long ago they may have wandered the windswept desert dunes as loners and rarely encountered the opposite sex. Of course, the camel is so totally dependent on humans now for its reproductive experiences that it's hard to know why it originally embraced the egg-on-demand strategy.

Thus evolution has allowed the females of some species to adjust the phase sequences of their physiological cycle and the timing of ovulation, for example, to accommodate a short life span or a solitary life-style. But most female mammals are restricted to regular cycles and can not place their eggs on hold until they meet a male.

Those physiological accommodations to extend egg availability are special cases. Since the life expectancy of an egg is brief, the egg self-destructs, usually within twenty-four hours after release from the follicle cells, and the female conforms her behavior to meet the needs of this ephemeral yet dictatorial gamete.

The time of reproduction must revolve around the time of egg production. From the hub of the egg radiates all the behaviors and physiologies that so patently create the set of strategies describing a female of any species more eloquently than the simple definition "a bearer of eggs." The female mammal announces the availability of mature eggs, turning from a sedate, calm, disinterested individual into one who pursues a male with passion exuding from the top of her head to the bottom of her rump. Eggless, this very same female cares not for a male, and her thoughts turn away from sex with an indifference that makes the human concept of sexual frigidity seem like a spring thaw. Most females seek sexual engagement only during the time that their physiological and behavioral cycles coincide to produce the egg and the estrus, and could not care less about it at any other time.

Being Female Without Reproducing

In most species the female produces eggs during her entire lifetime. Other than the human, no species faces a post-reproductive stage as a typical part of life and none has the option of not reproducing. Humans, however, control the reproduction of other species — the domesticated animals — with two opposite goals in mind; to curtail it, on the one hand, and to increase fecundity, on the other. Curtailing it permanently by gonadectomy is easy and has been done for hundreds of years. Curtailing it temporarily is more difficult and has led to such ingenious devices as cervical caps. Enhancing fertility and fecundity is much more difficult, for it entails tampering with the physiological system, which is complicated to manipulate and to keep functioning properly. To enhance fertility, humans take advantage of mammalian cycling to pinpoint the proper time for artificial insemination and, more recently, have taken advantage of the same cycling to implant embryos of one female into the uterus of another.

Humans have had much less success in manipulating their own birth rates, although not for want of effort or lack of imagination. In the absence of modern contraceptive methods, conception has been "prevented" by means of such bizarre prescriptions as hanging weasel feet around a woman's neck, stepping three times over a fresh grave, making cervical plugs out of chopped grass, douching with a well-shaken bottle of Coca-Cola, packing crocodile dung and honey over the cervix, spitting into a frog's mouth, and eating bees. Such contraceptive mechanisms have mixed results and most certainly will not replace the pill. Only during the twentieth century have methods of birth control become sufficiently reliable that a woman who chooses not to have children does not have to.

Along with antifertility drugs and techniques came the discovery of methods to improve fertility of women who experience difficulties in conceiving. These modern methods replaced older ones, which, albeit not as bizarre as primitive contraceptive prescriptions, were just as ineffective. Such rituals as picking fruit, drinking brews made from wasps' nests, eating spiders' eggs, and making and caring for baby dolls were some of the more popular choices. These fertility rituals evoked the fecundity of nature, with the hope that some of this fecundity would rub off on the women performing them.

Although some modern cultural institutions still decry antipregnancy methods, many Westernized women choose to delay the onset of reproduction — by way of the pill, intrauterine devices, or diaphragms, until they reach their twenties. Then they space their children over a period of several years and end reproduction by the time they are thirty-five. They have reduced their forty-five-year childbearing period to fifteen years.

In effect, the Westernized woman has created a cultural menopause. Despite the fact that she continues monthly menstruation as a member of the species, she has made herself as infertile as women whose reproduction has been ended by natural biological menopause. Although a woman may make the choice of not reproducing after age thirty-five, she is free to change her mind — perhaps because of a new marriage — and if she wishes, she can still bear a child. She is in charge and can reverse her cultural menopause.

Birth control gives the Westernized woman psychological freedom, because if she had to worry constantly about unwanted pregnancies, as did women in the past, she would undoubtedly become anxiety-ridden (perhaps even fearful) about the sexual encounters of marriage. As her age increased, she would have resorted finally to abstinence, as did many of her female forebears. Even if she created a menopause earlier, the Westernized woman often reacts to biological menopause with distress, depression, and a sense of loss — not of her

TIMING IS EVERYTHING

fertility, but rather of her control over the choice to have offspring or not.

Menopause is the gradual cessation of menstruation, which accompanies a decline of ovarian estrogen production. For reasons as yet unknown, the ovary, still containing eggs, puts out fewer and fewer estrogen molecules. Estrogen normally controls and reduces the pituitary's output of FSH and LH. So, since it is a feedback system, the decrease in estrogen leads to a dramatic rise in FSH and LH production. Indeed, imminent menopause is measured by the increased levels of these hormones in the blood. Furthermore, since an ovarian follicle and a corpus luteum are not produced, the uterine lining remains static and does not undergo its cyclical buildup and disintegration — hence cessation of menses. In addition, other distinct symptoms of menopause are "hot flashes", vaginal dryness and depression.

When her reproduction is suppressed either by choice or by menopause, the uniqueness of the female human is emphasized. No other female mammal can ignore the behavioral dictates of a ripe egg nor can she ignore the impelling forces that drive her to get her egg fertilized.

5

IN THE HEAT OF ESTRUS

THE PRESENCE OF a ripe egg in the middle of the estrous cycle dictates that a female will wish to mate and that she will seek, solicit, and stimulate the opposite sex to mate with her. Estrus probably was first noticed by farmers who were concerned with breeding their domestic stock. Breeding was only successful if they allowed their females to mate during their estrus. Knowing when the female is in estrus is still essential to farmers who cannot overcome the basic biology of females (although they have that of the male, with the exception of his sperm).

For example, the modern pig farmer checks his young sow's sexual condition by stroking her back. If the young female sow remains still, he knows that she will soon be ready to receive the first attentions of a boar, but he doesn't need the whole boar. He needs only the boar's sexually stimulating pig scent to further arouse her. Knowing this, the farmer sprays boar odor from an aerosol can across the young sow's sensitive nostrils, massages her hams, and, as she stands taut enjoying the odors of the male and his apparent rubbing, the farmer plugs her vagina with boar semen. However, unless she is in estrus, she will not take to this manhandling, darting away from intrusion into her privacy as quickly as possible.

Knowledge of estrus has not been confined to farmers. Some of the more stunning behaviors have been described and tallied by hundreds of scientists. For example, Frank Beach, a well-known comparative psychologist at the University of California at Berkeley, watched female beagles in heat as they actively solicited males. Beach recognized the great efforts that the bitch goes to to assure egg fertility. She does not retire coyly to her den and await the opportunistic visit of a male who may have sniffed out her estral aromas. Instead she goes on the prowl, seeking out the company of males and leaving a trail of telltale scents so that they may find her. Beach classified this activity as proceptive behavior.

During estrus, the female's proceptive behavior is designed to sexually arouse the male. Depending on the species, a female mammal may waft subtle perfumes, smack her lips at the male, bite him, knock him over, mount him, or mouth his penis, to name but a few strategies. The lowland gorilla female, for example, reverses the King Kong macho stereotype. The male is neither aggressive nor imposing, but actually rather unobtrusive. Ronald Nadler reported that the female makes the first overture, is "very assertive, backing forcefully into the male, frequently pushing him against a wall, and actively rubbing her genitalia against the male by rhythmically raising and lowering her rump while emitting a soft, high-pitched fluttering vocalization." The female initiates copulation as she slowly and seductively approaches the male, giving him the once-over, and if she likes what she sees — that is, a male who raises his arms as if to embrace her — she turns her back to him and presents. She advertises her ovulation with these behavioral changes, repeating them with the same male about every three hours. Though her copulatory activities are limited to estrus, she nonetheless copulates during pregnancy, through which her cycles continue. Pregnancy does not suspend estrus in the gorilla.

A young female orangutan in estrus willingly gives up her habitual solitude to be with a male. She approaches, grooms, touches the male, and mouths his genitals, soliciting him to copulate. One young female, coincidentally named Lolita by researchers, was particularly determined. Biruté Galdikas-Brindamour, a primatologist at the University of California at Los Angeles, observed: "For more than three days, the adolescent female pursued the unenthusiastic adult male to initiate mating."

Chimpanzees ovulate only one day of their thirty-six-day cycle, but they conspicuously inflate their genitalia for about ten days prior to ovulation. Yet the female pursues the male for some twenty-four days, with increasing fervor as her genitals engorge to flag her approaching ovulation.

She copulates during that entire time, and in peak estrus she copulates at least once every three hours and sometimes more often, spending her whole day consorting with one or another male (the night is for sleeping). Actually, she spends very little of her time in the sexual act itself—only about sixty-seven seconds. Indeed, each copulation is barely more than a sneeze. The male takes one minute for courting and foreplay and seven seconds for intromission and ejaculation.

If we tally up her copulatory time, given ten copulations a day, she has invested eleven minutes out of twenty-four hours. Her sexual activities are so casual that they remind us of our typical greeting, the handshake.

The female rhesus monkey—the most extensively studied of all primates, barring ourselves—has had every stage of her sex life dissected, examined, tallied, and even analyzed in chemical retorts. Her face, nipples, and perineum are marked with a conspicuous shade of red, and the skin around her genitals swells. She approaches eager males and wafts her perfume past their nostrils. When presenting, she turns her rear toward the male and swings her tail to one side. She energetically bobs and sways her head, which must somehow serve to further arouse the already stimulated male. Her final touch is grooming. She never engages in group sex but pairs off with one male at a time, copulating with several in succession.

The domesticated ewe in estrus rubs her neck against the ram's body and noses his genitalia. While in heat, the ewe does not even mind having her groin pawed by the male's hoof and even seems to enjoy it, adopting the mating stance and wagging her tail. When she is not in heat, she will not tolerate a male's hoof on her hindquarters.

An elephant cow in estrus uses the brushy tip of her tail as an atomizer to mist her body with her own sexy perfume. She sprays urine onto this brush, draws it up past her vagina, and, with a loud slap, flings it onto her back, scattering droplets of aphrodisiac liquid over her skin.

That the behavioral changes of estrus, singularly dramatic and surprisingly brief, are directly in concordance with hormonal levels has been shown time and time again by numerous experiments. The experimenters restored hormones to ovariectomized females and created estrus anew. For example, ovariectomized ewes showed no interest in tethered males until treated with the ovarian hormones progesterone and estrogen. In addition, scientists implanted hormones directly into the brain and created sexually receptive females. Experiments on and evidence from rats, guinea pigs, heifers, sows, cats, bitches, ewes, and monkeys are so great that we accept as scientific dogma the idea that sex hormones create certain features of sex behavior.

Hormones modify sensory and perceptual mechanisms; they change not only the stimuli to which a female responds but also her sensitivity levels, enchancing them in some cases and depressing them in others. Or so it is thought. The ways in which the hormones modify these many systems remain largely mysterious. Much of the scientific evidence is equivocal and requires retesting under more rigorous experimental conditions.

Speculations About Our Loss of Estrus

Despite uncertainty about the mechanism, an observer cannot and does not miss a female mammal in heat. Her egg availability is announced in no uncertain terms. In addition to the behavioral drama, some females paint their sexual region in gaudy colors; others engorge the vulval areas, enlarging them and making them more conspicuous; some emit pheromones from the vagina and disperse them by tail swishing; others perfume their urine and spray it throughout the neighborhood. In addition, they actively seek the company of males, solicit them, and arouse them in a forthright manner. The female confines these behaviors to the times of ovulation, behaving quite differently at all other times and often, when not in estrus, is downright nasty toward males. All mammals advertise their ovulation somehow, with the exception of one species—our own. We conceal it.

Why the female human does not announce behaviorally her time of egg release but rather "conceals" her ovulation—doesn't paint her bottom red, doesn't spread bodily perfumes around, and doesn't show proceptive behavior—is a question that has confounded the most scholarly writers who spew reams of speculation about the human condition into the pages of scientific journals. Obviously, sexual heat has an important place in evolutionary history; it is vital to reproductive success and species future. Without it, most mammals would not reproduce. Why then should estrus have been abandoned by the female human?

Because we have no scientific evidence and no experimental animals that show the same patterns, the answer to that question has given rise to a free-for-all in which scientists call upon bits and pieces of evidence from other mammals and assemble them into a completed jigsaw puzzle. Different scientists rearrange the same puzzle pieces to fit their own theories about why women have no estrus.

During what we could call the Desmond Morris *Naked Ape* era of speculation on the whys and wherefores of human evolution in the 1960s and early 1970s, scientists had a somewhat fuzzy view of estrus. Lack of estrus in humans was thought of as more or less synonymous with continual receptivity, to parallel the continual sperm production and virility in males. The cryptic nature of human ovulation was assumed to be a by-product of an evolutionary selection favoring increased receptivity. Continual receptivity was, in the arguments of these researchers, necessary for the evolution of human society. Some thought it strengthened the bond between mated male and female and so was responsible for human monogamy and shared parenting; others thought it reduced competition among males for females in heat and so contributed to the remarkable ability of humans to cooperate, for example, as a group of hunters. Certainly, human society would be very different if women came into heat accompanied by irresistibly seductive musks and paints; if men, unable to ignore the stimuli of estrus, left their nonestrous, unreceptive spouses to pant around a neighboring doorstep like dogs desperate for a chance at a bitch in heat.

These early speculators—most of them eminent anthropologists and primatologists—were so eager to correlate uniquely human characteristics with monogamous pairing in a social setting that they overlooked the testimony of our primate relatives. And our relatives tell us several things. Loss of estrus is not a prerequisite to continual receptivity, and continual receptivity and concealed ovulation are not inseparable. Many primates have either long periods of receptivity and yet still show estrus, or they don't advertise ovulation but are receptive only during the time they ovulate. Humans are the only ones who combine continual receptivity with loss of estrus.

Some theorists postulated that long-term pairing depended on lots of sexual interaction. Again, some of our primate relatives, such as the monogamous gibbons remain together with the same mate yet copulate only around the time of ovulation. Those primates who exhibit long periods of sexual receptivity, such as rhesus monkeys and chimpanzees, are highly social animals who dramatically advertise their receptivity at estrus with physical signs that can't be missed by even the most doddering, senile male. Even though these females are receptive outside of estrus, their interest in the male lacks the ardor they exhibit when in estrus. As we mentioned earlier, a female chimp's perineum is grossly swollen for a third of her thirty-two-day menstrual cycle, and it is only during that time that she is frantically obsessed with sex, mating on the average of once every three hours over the entire ten days. She is also receptive, although less intensely so, for a few days preceding and following the ten days of peak swelling, even though she ovulates only on the last day or so of that period. All this sexual activity has nothing to do with maintaining long-term pair bonds, for the female chimp is among the most profoundly promiscuous of creatures, often mating with a series of different males, one right after another, most of whom wait in line for her attentions with little overt aggression or competition. And even when she does form a particular attachment to one male, consorting with him for a day or two, leaving the group with him to go "on safari" in the jungle, their sexual relationship ends along with that estrus. In this species, at least, virtually continual receptivity reflects an open social existence. The earlier assumption that human monogamy rose in response to long-term pair bonding and concealed ovulation is not supported by animal studies.

At this point, the discipline called sociobiology entered the scene. No longer could speculations about the evolution of human behaviors be fuzzy and vague, couched in general terms of intuitive reasoning. Now, decreed sociobiology, all speculation must be closely and carefully reasoned, focused on this one question: How did this behavioral trait benefit the bearer and her genes — that is, improve her survival and reproductive success?

The year 1979 witnessed a positive spate of speculation on reasons for our lack of estrus, since everyone simultaneously realized that none of the early explanations were satisfactory under the empirical scrutiny of sociobiology. Three fascinating discussions on this topic entered the literature that year — each one closely reasoned; each one chock-full of evidence from humankind and primatedom; each one determined to argue cogently the genetic advantage bestowed by the trait; and each one, starting from similar premises and using similar evidence, reaching a drastically different conclusion from the rest.

The first of these three scholarly works, by University of Michigan scientist Richard D. Alexander and University of Wisconsin scientist Katherine M. Noonan, presents an argument superficially similar to those of the *Naked Ape* speculators, but it is filled with the dark specter of conflicting interests, trickery, and self-deception in the innocent, Edenlike world of our protohuman ancestors. Alexander and Noonan argue that, in an evolutionary sense, the two sexes have different objectives in life. A female mammal, who must expend a lot of time and energy in bringing up baby, will do best by her offspring if she gets some outside assistance—the ideal being a full-time mate to cater only to his one spouse and the kids. The male, on the other hand, will leave more of his genes to posterity if he spends his time sowing wild oats instead of being tied down to one household. Nothing new here; this is a standard theorem of sociobiology, and there is evidence that, indeed, in both human and animal societies that offer a choice, females who have hubby all to themselves have more surviving offspring than their sisters in harems, while males with harems leave more offspring than their monogamous confreres. An anonymous ditty of long ago sums up this sociobiological tenet of conflict between the sexes:

Hogamous higamous
Men are polygamous
Higamous hogamous
Women monogamous.

The authors then startle us with their conclusion: "Protowoman" tricked "protoman" into being her perpetual consort and helpmate by concealing from him any clues to her physiological state, hiding any sign of ovulation.

Their reasoning runs as follows: Females who showed no external signs of ovulation would probably not be successfully impregnated by males interested only in one-night stands. A male willing to pair with such a female for a long time, on the other hand, mating with her throughout her monthly cycles, would have the best chance of becoming a father. Of course, these long-term attempts to impregnate her would leave him no time to chase other females, but then neither would he have to fend off other males interested in his spouse; they'd be paired off, too, and wouldn't be attracted to our heroine when she did ovulate, because they wouldn't have the slightest idea when that was. When the babies came, our hero would be as certain as he could be that he was indeed their daddy, so he wouldn't hesitate to feed and clothe them and put them through protocollege.

The point of concealing ovulation from males, according to Alexander and Noonan, is to force them into permanent wedlock. The female would not want to give any clues to her reproductive status. And so the authors suggest that self-deception — the suppression of the female's own awareness of her ovulation — aided her in fooling males, for she could not then betray her state by any revealing actions or gestures. Alexander and Noonan end their essay with these words:

"The ability to deceive, partly by self-deception as to motives, we here suggest to be a central part of human sociality, and of consciousness and self-awareness in the human individual. Concealed ovulation we view as a particularly powerful and instructive case of deception of others, linked with self-deception and made more effective by it."

Sobering thoughts indeed, and thoughts not unique to these authors. Another article (published soon after Alexander and Noonan's) by Lee Benshoof and Randy Thornhill, zoologists at the University of New Mexico, agrees about the deceptive power of concealed ovulation but reaches a profoundly different conclusion on nearly every other point, painting an even grimmer picture of the intrigue-ridden lives of our hominid predecessors.

Benshoof and Thornhill argue that, contrary to the reasoning expounded by Alexander and Noonan, concealed ovulation did not lead to the establishment of monogamous matings. They point out that males of many other species are content to lead monogamous lives because it is in their interest to do so — for example, because the offspring require more care than one parent alone can give, as is the case of the human infant — and not because they must be coerced or tricked by the female into doing so. Furthermore, these authors take exception to the idea that concealed ovulation increases the male's confidence in his paternity. They imagine the poor "protoman" constantly guarding his wife against covetous neighbors, in a lather of indecision about whether he should join the hunting party or stay home on the off-chance that this will be her fertile week. On the contrary, Benshoof and Thornhill say, no system could be more poorly designed to insure paternity than the system of concealed ovulation. If you want assurance that a particular male is the father of a particular baby, you must have what the other monogamous primates do: a short, well-defined estrus with very little advertisement of the fact. That way, the female's mate knows when she is in heat, but other males don't. Her mate need only guard her for a few days, and then he's free to go hunting, swap tall tales, or play golf, secure in the knowledge that however long he is gone from his mate, she could have been impregnated only by him.

These authors hypothesize that this kind of system—monogamy with estrus—came first in human evolution and that only later did estrus disappear—and for a reason diametrically opposed to that given by Alexander and Noonan. Benshoof and Thornhill believe that concealed ovulation allowed females to successfully cuckold their mates—to trick them into caring and providing for offspring who were not theirs.

The rationale for such behavior is grounded, they believe, in the social structure of humans. Ever since Desmond Morris's *Naked Ape*, scientists have recognized that the form of human society—monogamous pairings in a group setting—is unique among mammals. Monogamy is relatively rare in our class of animals, but those mammals who prefer it are almost always territorial, too. Living in groups, the experts argue, gave "protowoman" more males to choose from than her territorial counterpart, and it also made for some sociobiological conflicts. On the one hand, she needs a steady mate to help her rear her kids, but on the other hand, how could she possibly prefer (in the evolutionary sense) to make babies with the genes of her own ninety-eight-pound weakling of a hubby than with the genes of the hunk who lives next-door?

How does she reconcile these conflicting pressures? Simple. She just conceals her estrus, making love with hubby most of the time but sneaking over to be unfaithful with the hunk only when she's ready to ovulate. Hubby won't know that she wasn't fertile during any of the times she had intercourse with him, so when the babies come, he'll think they're his — assuming that "protoman" wasn't aware of the genetics of eye color or hair tint. His unfaithful wife will have garnered the best of both worlds: a doting "daddy" and superior paternal genes. What more could the kiddies need?

Analysis of the costs and benefits of cuckoldry is mainstream sociobiology. In fact, in 1981, sociobiologist Sarah Blaffer Hrdy of Harvard University proposed in her book, *The Woman That Never Evolved*, a very similar explanation for concealed ovulation, with the added fillip of practical polyandry. She proposed that cuckoldry would lead to more tangible gains than just better genes for baby: By mating with many males, females would form a web of male allies and acquaintances, each of whom might assist her and her offspring, since each might think himself the father.

The biggest difficulty with this cozy cuckoldry scenario is that women conceal ovulation from themselves as well as from their spouses and other males. Although some women think they can tell when they are ovulating, the vast majority most decidedly cannot, and even with our current technological ability to measure basal body temperature and to sample and categorize cervical mucus, the time of ovulation is notoriously difficult to pinpoint. (Witness the average time of four months it takes a woman to get pregnant when she and her partner are trying, or the legendary unreliability of the rhythm method of birth control.) Benshoof and Thornhill suggest that the female may have some unconscious physiological control over timing of ovulation, akin to the eggs-on-demand ovulators, but at best this would only work within narrow limits. Their theory does not explain why concealment from self evolved. Surely their suggested system would function infinitely better if the female herself knew when an egg was ready and confined her affaires de coeur only to that time. These authors must fall back on that old adage, "The best deceiver is a self-deceiver," to explain the female's unawareness of ovulation. But this explanation seems a weak conclusion to such a dramatic story, for the female can only deceive herself about when she was unfaithful to her mate, not about whether she was unfaithful.

A third explanation attempts to fill the self-deception gap. Nancy Burley at the University of Illinois in Urbana focused on the inability of women to know themselves, at least when it comes to ovulation time, instead of their ability to hide ovulation from others. Not satisfied with the speculations of others, convinced that they were inadequate to explain the phenomenon of concealment from self, she formulated a very straightforward hypothesis: If women knew when they ovulated, our species would have become extinct.

Burley argues that "protowoman," intelligent and creative enough to imagine the future, and — alone among the females of the world — self-aware enough to see the consequences of her actions, would have desired to limit the number of children she bore. If she was aware of her own fertile period, she might have deliberately refrained from sexual activity at that time, knowing that if she did not, she would pay the penalty of another pregnancy. Burley suggests that a woman's ability to sense her own ovulation would have led to effective birth control (a rhythm method that really worked) and hence to small families — perhaps too small, in the light of high infant mortality, to keep the species going.

Before you protest that women as a whole could never have been so antifamily as to not want children, consider for a moment what motherhood meant for one of our female ancestors. She mated at the ripe old age of fifteen, let's say, and between then and her death at a youthful thirty-five or so, she got pregnant and had babies on the average of once every three years or so — assuming that the infants lived long enough to suckle for a couple of years, and that she was lucky. If not, the pregnancies came even faster. So, in a twenty-year segment of her life, reductionistically named by actuaries the "childbearing years," our female lived through, at the very least, six to seven pregnancies, childbirths, and child-rearings. Other females weren't as fortunate; they didn't live through childbirth.

This is a picture bleak enough to make all but the most steadfastly maternal among us flinch, and yet this was the fate that awaited most married women up until the advent of modern birth-control methods. Burley cites a number of anthropological studies showing that, universally, women want smaller families than their husbands do (or their other relatives, for that matter), but there are few parts of the world where women actually have much control over whether or not they get pregnant. If women knew with certainty when they were fertile, we would be saved all the worry about the side effects of the pill and the IUD, about the social impact of abortions and unwanted pregnancies — that is, if our ancestors had refrained from practicing birth control to the point of extinction.

However, Burley may be underestimating the power of human

culture and society to counteract our most deep-seated evolutionary desires and even to convince us, with an efficiency that puts Madison Avenue to shame, that we want and need something we quite definitely do not want or need. Historian Lawrence Stone in an article in the *New York Times Book Review* sums up what in all probability was the attitude of women toward maternity in traditional society: "Although women feared childbirth, they were taught to regard it as their natural duty and accordingly were proud of it. Indeed, some scholars argue that menopause was psychologically more traumatic in the past than today just because it deprived women of what they regarded as their prime purpose in life, reproduction."

We'll never know which of these hypotheses, if any, is closest to the truth, or if bits and pieces of several combine to describe what forces shaped early woman. The scientists disagree on much, but not all of what they say is mutually exclusive. They do agree, though, that what needs to be explained is the phenomenon of concealed ovulation, rather than that old warhorse, continual receptivity. In fact, most recent speculations agree that if there is anything woman is *not*, it is continually receptive. As Alexander and Noonan point out, in comparison with the overwhelming sexual drive of female mammals in heat, "the human female's behavior might best be described as a kind of selective or low-key receptivity, commonly tuned to a single male." For those other males, to whom she gives the cold shoulder, our female human is much better described as "continuously nonreceptive." As University of California psychologist Frank Beach succinctly and wittily put it, 'No human female is 'constantly receptive' (any male who entertains this illusion must be a very old man with a short memory or a very young man due for a bitter disappointment)."

No matter what various reasons they propose for the concealment of ovulation, these speculators do agree that women lost the cyclical receptivity that characterizes estrus, without losing the cycle itself, because the time of ovulation had to be hidden.

Estrus: Biological Excess Baggage?

The temptation to speculate is a strong one, and so we'll jump in, feet first, with our own unprovable and untestable hypothesis. We claim that the term *concealed ovulation* begs for an explanation rife with deceit and intrigue. The connotations of the word *concealment* have influenced, we think, the way that scientists approached the explanation for our lack of estrus. If time of ovulation was hidden (so would go the argument), it had to be disadvantageous to reveal. Consider, for a moment, a different term: *loss* of estrus. This term, without the overtones of the word *concealment*, engenders a different train of thought. Need there be a strong, gripping reason why estrus had to be jettisoned? Not necessarily. Evolution is always chucking out unwanted bits of biological baggage; structures and behaviors are always getting "lost," not because they are liabilities per se, but because they are dead weight; they just aren't needed anymore.

For example, lots of cave-dwelling critters—fishes, insects, amphibians—don't have eyes, because they don't need them; it's energetically expensive to build eyes that you can't use. Even some features that *we* could use now were eliminated by evolution because in the past they were excess baggage, of no use—the ability to synthesize vitamin C is one. Of all mammals, only guinea pigs and primates lack the ability to manufacture their own supply of this essential vitamin and must rely on diet to supply it. If your diet consists of nothing but raw fruits and vegetables, you don't need to make it, because you eat plenty of it. You only miss it if you have become an omnivore (as we have) who cooks or preserves food (as we do).

We suggest that estrus was lost not because it had to be concealed, but because, like troglodyte eyes or guinea pig vitamin C synthesis, it was not needed anymore. Many species of monkeys and apes, although they display estrus as other mammals do, have managed to partially disconnect receptivity from the hormonal highs of ovulation. The female chimp, as mentioned earlier, will copulate over as much as two-thirds of her menstrual cycle. Some rhesus monkey females apparently seek out and mate with males irregularly throughout their cycles. Even female gorillas, who usually confine their sexual activities to a short estrus coinciding with ovulation, commonly display one or more anovulatory estrus periods during pregnancy, when they are as aggressively proceptive and seductive to the blasé male as they are during any estrus when they do ovulate.

For primates, sex is apparently not something to do just to make babies. For many primate females, it seems to have become a casual interaction—social intercourse, if you will—more akin to mutual ectoparasite-picking than to the frenzied sexual drive of estrous females in other groups.

We suggest that estrus in women followed this primate pattern of diffusion over time, but that its loss, unique among mammals, was the product of another unique human attribute: living in a culture that pokes into every corner of our lives. Perhaps societal expectations decreed that women should accommodate their husbands, "doing their wifely duties" at any time of their menstrual cycle—on their marriage night, before the hunt started, or after its completion, no matter what her physiology told her to do at these times. From the standpoint of evolution, the easiest way to do this was to "switch off" estrus, to no longer make women "receptive" in terms of the frenzied burst of sexual drive characterizing heat, and to substitute a low-key accommodation, a sex drive with little variation over her cycle—the condition that leads to the paradoxical description of women as both "continually receptive" and "continually nonreceptive." Ovulation wasn't deliberately concealed by evolution, but its signals became secondary, unimportant ones that, if they did anything at all, probably interfered with the pattern of life in "protohumans," where rite and ritual, rather than the ebb and flow of one's hormones, may have determined when to make love and when to be celibate.

As the female cycle became less important in determining when husband and wife would copulate, a reversal of roles occurred. The female relinquished her position as the arbiter of sexual readiness, and the male took the initiative. No longer did women, like girl gorillas, present their rumps to an unaroused male or squash his mistimed interest with a curt refusal. Now man became the demander and woman the slowly aroused. This role reversal took hold so completely that the perpetually uninterested wife is a cliché of modern society, and that guru of the sexual revolution, Helen Gurley Brown, feels it incumbent upon her, even in the 1980s, to berate wives who plead nocturnal headaches.

But not all female initiative is gone. Since receptivity in women is mainly under the control of the brain—rather than the gonads, as it is in females with estrus—women are free, within the limits of their own psychology and the decrees of society, to choose the times and places of sexual contact. Far from being slaves to their raging hormones, automatically echoing the trends of a hormonal roller coaster

IN THE HEAT OF ESTRUS

with behavioral ups and downs, women show remarkably little divergence from a steady level of sexual activity throughout their cycles. In fact, the human female could pass very well as a creature without an ovulatory cycle, except for the obvious monthly event of menstruation.

Whatever the evolutionary reasons for human loss of estrus, the results of this loss included the development of love as a corollary to human sexuality. As the act of copulation became more and more independent of the hormonal highs of ovulation, humans began to copulate for other reasons—for companionship, for amusement, for power, for love. Less and less was fertilization of the egg the goal of copulation for the female human who did not advertise the presence of her ripe egg by a change of behavior or by a change in how she looked. Some religions, such as Roman Catholicism, insist that the function of human copulation is solely for procreation, but by that insistence they are unwittingly returning humans to the level of other mammals and equating human sexual drive with the frenzy of estrus. Humans have evolved a *different* sexuality from all other mammals, and neither science nor religion should deny that here humans are unique.

6

A Million Ways to Catch a Male

THE EGG IS READY, ripe, chock-full of nutrients, but like Cinderella on the night of the ball, its moment of glory is brief. The female with ripe eggs is desperate. The clock is her enemy, aging the egg, ticking its brief viability mercilessly away. Burdened with the transitory nature of eggs — unless they are fertilized, they will die — she is impelled, at all costs, to find a suitable sperm resource. If her eggs are not fertilized, she loses in the biological race of personal success.

She is under far greater pressure to mate at that time than the male, who may mate successfully at any given time. Among mammals, for example, even during breeding seasons eggs are produced episodically, while sperm are produced continually. Thus, if a male doesn't mate on a particular day, there is always the next day. The female, on the other hand, is pressed to mate, and if she misses that day, she may not have another opportunity for an entire year. The species future rests on her; consequently, it would seem logical that she is the driven one, the one who pursues. Yet the stereotype of animal courtship and mating always casts the male as pursuer, the one who must cajole, bully, and seduce the coy female. The male is seen as the stud, attempting to service as many females as possible, and is seen as having only one thing on his mind: copulation. Observers, blinded by the shenanigans of courting males, fail to see the techniques the female employs to satisfy her sexual appetites, to get her eggs fertilized. The male's gaudy appearances and behavior are not intended to dominate the female but rather to make him obvious and conveniently located.

Males produce sperm continuously during a breeding season as a convenience to the female; they provide a sperm resource for females whose eggs are readied only occasionally. Females of many species share the need to find males and to get sperm from them in whatever manner their biology demands. Although males are basically a sperm resource, females—clever sex that they are—use males in other ways as well. How a female makes best use of a male depends on her needs to fulfill her reproductive mandate, which varies from species to species, depending on the constraints of her life-style.

If she is a sessile (immobile) female, such as a clam, oyster, or mussel, she can't do very much, but neither can the male. A sessile female can't shake off the mud and sand from her bivalves and clamber around to court clusters of male bedfellows. Instead, as a way of getting together with her mate, she synchronizes her biological time clock, as does the male, to phases of the moon, to tides, to temperatures, and after releasing her eggs, she can only "hope for the best." The precision of egg and sperm release is carried to rare perfection by a sessile segmented worm, which literally detaches itself from its sexual experiences. Adults of the palolo worm, restricted to cryptic burrows among the coral reefs of the South Pacific, don't ever venture out of these protective rocky fortresses for anything, not even to mate. Instead, as the changing day length signals the advent of the October breeding season, they grow special segments on their bodies that carry all their ovaries or testes. Their timing is so precise that on the night before the last quarter of the moon, these rear body parts break off from every palolo worm and become swimming gonads. The ocean seethes with thousands of pulsating, headless segments, all attracted to the moonlight.

In the seething stew of thousands of upwardly mobile gametes, sperm find eggs and fertilize them, and the zygotes sink downward to rest on the bottom, where they develop into the next generation. Meanwhile, the adults renew themselves and take the pause that refreshes. After the rest, they regenerate new sexual segments to replace the jettisoned ones, in readiness for the next lunar titillation.

Other sessile females use other tricks to get their eggs fertilized—the abalone waits for sperm to swim by before releasing eggs; the oyster sucks sperm into her body so that the eggs will be fertilized internally; and the sea urchin sends her eggs out enveloped in a cloud of chemical stimulants, prodding males to release sperm at that moment.

Life within a shell is certainly very confining, and under such con-

A MILLION WAYS TO CATCH A MALE

ditions the female does the best she can to get her eggs fertilized. Since she is rock- or mudbound, she sends her eggs out by the millions, and through saturation of the seas (at the right moment, of course), she guarantees that at least some of her eggs will get fertilized. Her strategy works because the male is doing exactly the same thing, releasing billions of sperm in the same fashion. Sessile species have no courtship, and if they did, the female would be hard-pressed to identify a suitable male bivalve.

Here is true gender equality: They look alike, they feed the same way, and they react identically to surrounding stimuli. Uniformity is the key to survival; if an idiosyncratic oyster's clock ran too fast or too slow, she would not leave any offspring.

Sessile spawners are immobilized, frozen into space, by the hand of evolution. But when water-dwelling females become unstuck, a whole new array of strategies for getting their eggs fertilized evolve. Instead of having to cast her progeny into the vast unknown ocean to meet their fate in the gullet of a fish, this female can travel to a place where her offspring will find a tender, loving environment. Females of some species migrate over long distances to spawn in special nursery areas, such as the fjords of Norway or the bays of San Francisco, drawn there by the particular needs of their eggs. They are accompanied on their trek by the males with whom they will pair but briefly.

The first home of a Pacific herring, for example, is an underwater grassy bed selected at the end of the winter by its mother. Before she swims to the spawning grounds, she has spent many months swimming in vast schools in the ocean. Stimulated by the changing season, she and all her schooling cohorts migrate en masse to the shallow, weedy bays. The sight of the abundant vegetation and the smell of other herring spawning overtakes all the newcomers, who join in the mass spawning orgy. Evidently, by rubbing the right weeds, the female is stimulated to lay eggs. She turns on her side, extends her fins for balance, arches her back, and brushes her egg-laying tube across the strands of grass. The eggs, as they are extruded, stick to the grass, and any nearby male is welcome to blanket the newly laid string of eggs with his sperm. But it doesn't matter whether or not a male is close by. She is surrounded by clouds of milky white sperm shed by hundreds of thousands of fish all spawning simultaneously in the nursery area.

The female herring mates on the spawning grounds, for there her eggs have the best chance of developing, at least during the first ten days of their life. Any male or males serve as a sperm resource, their identity immaterial to the female's concerns. The spawning site where thousands of fish slip and slide past one another as they shed gametes becomes a huge sperm stew, and all eggs are likely to be fertilized.

Female squid engage more actively in pairing with males. Like the herring, opalescent squid normally travel in schools, although their sociability is limited to swimming in regular geometric configuration, without apparent notice of one another except to remain a certain distance apart. At mating time, however, drastic changes alter not only their usual geometry, but the female's indifference to the male as well. Primed by the lengthening days of spring, females begin to manufacture eggs and, accompanied by males, to migrate to a seabed along the coast of southern California. There in the newly dark dusk and the first glimmers of dawn, the female advertises her readiness to mate by projecting her squidy musks and exhibiting a coral-colored shell gland, easily seen through her translucent skin. But like the herring, she will not extrude her eggs unless she is in the right spawning area, as evidenced by ropy masses of squid eggs entwined with the algae on the sea floor. The sight of the egg masses is a potent aphrodisiac for both sexes, assuring the newcomers that they have found the right seabed for a one-night stand. The newly receptive female allows any nearby male to hug her in a many-armed embrace. While holding tight to his newfound love with most of his arms, the male inserts down his gullet a specialized arm, called the hectocotylus, and he grasps a suckerful of spermatophores. Armed with his spermatophores, he lifts out the hectocotylus, and the female permits him to plunge it into her sperm receptacle. Then the pair untangle their arms and separate. The female, now concerned with egg laying, passes a string of eggs out of her body, fertilizing them on the way. She cradles the newly fertilized eggs in her arms and, upon sighting an egg mass, she gently weaves her own rope of eggs firmly into its center. In effect, she is camouflaging her eggs by hiding them against the most perfectly matching background possible: the eggs of hundreds of other females. The end result is an incredible tangle of squid eggs covering an unimaginable area as large as forty square feet. The strings are interwoven in such a complex way that a predator would be unable to tear any string away or to swallow the entire huge mass. Furthermore, the sea's currents would have a hard time uprooting this vast tumbleweed of the squid's next generation.

The female, after laying, is 53 percent lighter and shows the ravages of being hugged by so many arms during her brief pairing. Bits of her skin are torn away, sores develop on her body, her normally bright eyes cloud over, and she seems to be tipsy. Her degradation lasts only a short while, for she soon dies, as does the similarly mutilated male.

For the squid female, the male is a mere reservoir of sperm; aside from fertilizing her eggs, he has little to do with stimulating spawning. Like birds that lay eggs only when the nest is finished, squid spawning is stimulated by the sight of other females' eggs entwined in ropy masses. In all probability, if a female did not see the reproductive product of her sisters, she would withhold her spawn until the environmental appearance matched the environmental pattern etched in her cephalopod brain. For the female squid, the male is not a sufficient stimulus for her to complete her reproductive mandate. She must rely on other females who tell her where to spawn, and by taking her cue from their consensus, she improves her offspring's odds for survival.

The herring and the squid live in huge schools, mainly to confound predators and to improve their feeding efficiency. But the society offered by schooling provides more than protection and better nutrition for these species. Their natural tendencies to live in huge groups serve their reproductive techniques as well. They use each other as stimulants for spawning. The presence of hundreds of species cohorts extruding gametes triggers a similar response among species members who have not begun to reproduce. Since members of a school look identical, are interchangeable, and have equal voices in the structure of the social group, it matters not which male fertilizes the eggs.

Members of a school are special. They epitomize equality: One female is exactly like another, one male is exactly like another, and aggression among school members is nonexistent. All school members behave in precisely the same way, often interchanging positions, swerving away from predators at the same time, and spawning synchronously as well. Schoolers give up personal space territoriality and individuality in exchange for the advantages provided by living in great organized groups. However, most species are not so socially harmonious. Schooling as a life-style and individuality as a life-style are anathema; it is not possible for the two to coexist.

If establishing a territory or home range is a species requirement, it brings with it the emergence of the individual. The need for personal space, which is often at a premium, places special demands on species members and leads to the creation of different roles; with the creation of roles, the uniqueness of the individual emerges. Unlike schooling males, one male is not exactly like another, and, for the female, a new freedom emerges—the freedom of choice. She can choose a male.

In a process known as courtship, prospective mates are selected by the female. The female's selection—assuming that she wants the species' best—places a huge burden on the male. He must meet her criteria and he must strive to excel, to hold the best territory, to sing the loudest and most melodic song, or to grapple with other equally striving males, depending on his species. Behaviors such as these are the hallmarks of stereotypical courtship. From these courtship displays rose the ever-present notion that courtship behaviors, seemingly instigated by males, are the male prerogative. Not so. The male performs in response to the female's reproductive mandate and her autocratic egg; to be successful, to gain acceptance, to mate, a male must please the female.

Males of different species must meet different criteria. In some, the male only need be a virile sperm resource. In others, he may have to be a potent physiological stimulant, a diligent protector of territory and the young, a generous provider and a compatible lifetime companion. Even though the courting display may be initiated by the male, sometimes without a female present, the courtship dialogue cannot begin until the female decides if she is going to accept the male's overtures.

The Lek: A Sexual Supermarket

Among the most strikingly theatrical courtship displays that starts in a sexual void is one that takes place in a lek, a breeding arena in which groups of males congregate solely for the purpose of allowing the female to look them over and choose the one with whom she will mate. Each male patrols a crowded little territory, crammed side by side next to his rivals.

Although uncommon, leks are characteristic of an odd assortment of species. The most spectacular and best known are the leks of sage-grouse males, who, during their spring breeding season, flock by the hundreds to traditional arenas on the prairies of Wyoming and Montana. Tradition defines the location of the site, and the birds return there year after year. The arena may cover the same two to three acres each breeding season, and some leks may have been in the same location since the glaciers receded to their current home in the north, thousands of years ago. The massed booming sounds of lekking male sage grouse can be heard for hundreds of yards, and their startlingly white bobbing chests can be seen a mile away. The congregation of males, so obvious to human observers, cannot be missed by sage-grouse females who have more acute vision and hearing.

Females migrating down from their wintering spas see and hear the males broadcasting their sexual readiness through their orchestrated noise and activity. What do the hens see as they fly in on a frosty early spring morning to Muddy Spring, Wyoming, a lekking site for at least the past forty years?

They see more than one hundred males expanding their ruffs, inflating their chest sacs, booming, strutting, spreading their feathers, and stomping around, each on his own piece of barren ground the size of an ordinary bedroom. Initially cautious, hens congregate around the edge of the arena, seemingly uninterested in the blaring love calls coming from the cocks. At the lek's periphery, the hens chatter among themselves, feed, groom, and preen their feathers in an unhurried manner. However, in a few days, blasé indifference to the hundreds of cocks booming out love calls, showing their manliness, and advertising their bulging sperm assets gives way to a blossoming reception. One morning the many hens awaken in a different mood.

On that special morning, a hen flies in alone in the first rays of the warm morning sun and strides purposefully into the lek, ignoring many alluring, enticing, and courting males as she moves toward the center. At the center, clusters of hens mill around near the central males and, upon seeing them, she joins the flock. The strutting and booming males court each female, who patiently awaits her turn to copulate with one of these central males. After her early-morning fling, the female wanders out of the lek into the surrounding sagebrush and disappears. The male has served her purposes in providing sperm to fertilize her eggs. Now she alone has the task of incubation.

But why does the female choose the male in the center when the males around the periphery would certainly be most pleased to provide sperm, if we can assume that their frantic antics indicate sexual readiness? Are the central males more handsome, louder boomers, or more seductive strutters? Apparently, she is not drawn to them by their skills in courting, since all the males at the lek appear to display themselves with equal passion. R. Haven Wiley, a zoologist from the University of North Carolina at Chapel Hill who is schooled in the ways of the sage grouse, reported that he could not distinguish differences in appearance or courting behavior between centrally located males preferred by females and peripherally located males, who were regularly overlooked. Furthermore, the quality of the male's central territory, a poor piece of real estate at best, is probably not the criterion for her choosing the male, since she doesn't feed there or raise her young there. And yet this central spot makes the male sexually desirable; the females will mate with any male who occupies it. Does the central male have a secret asset?

It turns out that the central males are fortunate creatures of circumstances. In the right place, they copulate with most of the females. However, their allure stems not from the place, but rather from a set of conditions that serve the female's needs.

Let's go back briefly to some of the biological problems that a flying female vertebrate, such as the bird, has to face. Her gametes are heavy, large, yolk-filled eggs, which, when matured, are one-third of her body weight, a hefty burden to lug around, especially at take-off time.

It makes no biological sense to fly around overloaded, and thus, a biologically smart female shouldn't mature her eggs until she nears laying time. Maturation may proceed at a rapid rate when she reaches the mating ground and is reassured that sperm abound. Speculating on the sequence of events, we think that hanging about the periphery of the lek is a strategy to ready the eggs for fertilization. The sight and sound of the males help to mature her eggs as well as arouse her sexually. When she finally enters the lek from the periphery, her sexual arousal is heightened by the superstimulating displays of gyrating, strutting males booming out their love calls. Walking toward the center, no matter where she turns her head, she sees displaying males and hears their sounds. She sees other females at the center, and she may need both their reassuring presence and the superabundant visual and auditory stimuli of many males to copulate with one.

Pondering about the function of a lek, from the male viewpoint, biologists have been unable to determine how it benefits males, especially since so few males have the opportunity of copulating. But, seen from the bird's-eye view of a sage-grouse hen, the lek serves a very important function: It is a physiological resource to prime her eggs; it serves to arouse her, and it provides trillions of sperm to fertilize her eggs. She can't miss being inseminated.

Mammalian leks serve the females differently. Few mammals lek. One that does is a medium-size antelope, the herd-dwelling Uganda kob, which inhabits the equatorial plains of Africa, an essentially seasonless land where temperature and day length remain fairly constant all year long. In the absence of seasons, the female is free to give birth at any time of the year, and consequently she can come into her one-day estrus at any time. But when she comes in heat, she can't conveniently solicit the attentions of a male neighbor, because there are none. She has excluded adult males from the herd, even rejecting her own adolescent sons. The all-female herd, with their juveniles, wander widely, grazing over the equatorial savannas. When a female comes in heat, she could be in grave trouble. She must find a male before her one-day estrus comes to an end. The male lek now serves her well. She trots off to a male lek, knowing just where to find it, since the location of the lek is traditional and constant, year in, year out. Although individual males may come and go, the geographic site stays the same.

Thus, if a female in heat finds herself in a particular place at the northwest corner of Uganda, she knows that a lek is just over the next hill; similarly, if she is in another familiar spot in the south, she knows that a lek is just to the left of that clump of trees. Hence, females in the peregrinating herds need spend little time in search of a male. They know the location of male homesteads when their sexual urge needs to be fulfilled.

Like the sage grouse, she wanders into the middle of the lek. Like the sage grouse, the presence of other females seems to reassure her that this is a choice spot. The female casts a discerning eye over the scene, and if she spots other females clustered near the center, she walks directly toward them and greets them. After the greeting, she feels comfortable enough to move among the males in the lek, eyeing them and smelling them without hindrance.

The males make no overt overtures, but they watch her carefully, waiting with baited breath to see who will be the chosen one. Stopping within a male's territory is her symbolic gesture, essentially informing the male that she is waiting to be seduced. Not to be misled, in case she is merely a flirt, the male checks her physiological condition by sniffing her vulva, which stimulates her to urinate. Almost gluing his nose to her vulva, he tolerates a splash of her pheromone-scented urine and seems to revel in the drenching his head receives, for then he raises his head and curls his lips and nostrils in a satisfied grimace. Then she permits the male to mount her several times in order to achieve an erection. After about fifteen minutes of love play, the exchange of ardor is consummated with copulation. The female may repeat this love play with the same male, or she may stroll into another's territory and splash his nose with her seductive urine. Helmut Buechner and Robert Schloeth, zoologists from Washington State University and the Swiss National Park, watched an elderly female seduce a number of males, permitting 219 mountings, which culminated in seventeen known ejaculations, all within a seven-hour period. Obviously, experienced females know how, with a minimum of exertion, to interest the males.

In contrast to the sage-grouse female, who uses the lek to help prepare her eggs as well as to get sperm, the Uganda kob female makes use of a lek solely as a sperm resource, since she arrives at the male's territory with her egg fully readied for fertilization.

The strategy of grouping reproductively ready males in a lek is not restricted just to vertebrates. Some species of a singularly successful genus, the pesky fruit fly, *Drosophila*, also lek. In Hawaii, fruit-fly females find a perfectly suitable home for themselves and their eggs in rotting vegetation, thriving in the midst of this warm fermenting mass. The female spends most of her life there, except when the urge to mate arouses her tiny body and impels her to wing to a nearby tree fern, where she is likely to find lekking males. The males buzz around, waggling their wings and swabbing their own territorial leaf with fluid secreted from their abdomens.

Attracted by the communal aroma of males, the female flies into the lek and selects a male, with whom she exchanges for a few minutes drosophila small talk as they dance the courtship tango. Copulation lasts for several seconds and then the female rests with the male, exhausted by her sexual encounter. Refreshed by her brief nap, she leaves the male behind and returns to her rotting vegetation home, which now becomes the nursery.

Drosophila species elsewhere don't form leks. Normally, copulation takes place right on the rotting vegetation and the female oviposits where she has had her sexual tryst.

Leks evolved, according to Herman Spieth, an entomologist at the University of California at Davis, because of the Hawaiian fruit fly's life-style. Rotting vegetation attracts both fruit flies and some Hawaiian birds. Birds like to eat fruit flies. Normally, the flies successfully avoid the birds' beaks, but there are moments in the life of a fruit fly when it is completely oblivious to avian threat. Like other animals intent on copulating, fruit flies may be so sexually aroused by their courting behavior that they become oblivious to dangerous predators lurking nearby. Since birds feed at the same troughs as *Drosophila*, a mating pair of insects intensely buzzing and dancing could well become the appetizer course for bug-hungry Hawaiian birds. Thus, it is thought that the fruit-fly lek probably originated as an antipredator device so that mating could take place away from the rotting vegetation without a feathered predator hanging over the copulating pair.

For most species, a lek makes no biological sense; it is a bizarre evolutionary phenomenon that removes courtship and copulation from everyday life. Using the lek primarily for copulation, a female neither feeds there nor lays her eggs within its generally impoverished, limited boundaries; and the male does nothing at a lek except provide stud service, and perhaps challenge other males for the desirable territory.

It's hard to repress the images that pop into mind when we compare an animal lek to the phenomenon of human males gathered around the polished oak and brass of a singles bar. Like a lek, a singles bar remains in the same geographic location day in and day out, and the same males do not necessarily rest their foot on the same bit of rail from one night to another. Also like the lek, a singles bar removes courtship from the routines of everyday life. Do the males go to these bars for the same purposes as the sage-grouse cock? Is a brief courtship followed by copulation that evening their ultimate goal? Are they willing to consider the possibility of a lasting relationship?

The singles-bar lek is an experiment in cultural evolution, and its prognosis for survival is poor. Although the lek serves the needs of some females who know the rules and enter it for a brief fling, a one-night stand, it brings with it a heretofore unexpected danger, which will probably lead to its extinction. The desire for casual sexual encounters has been undermined by the fear of contracting the incurable herpes virus, a devastating venereal disease. Cultural evolution will in time select against the singles bar and it, like the dinosaurs, will disappear from the scene.

When She Does the Asking

Females approaching a lek confront a panorama of males, all ready to mate, from whom to choose. As a superstimulating group, they are convenient, but finally, no matter how aroused she is, she can only mate with one male at a time. Females of most other species don't seem to need the overwhelming view of hundreds of males, parading and singing together in their sexual finery, to be aroused. They are content with a one-man show. And show it is, for the females watch as the males go to a great effort to attract. Males fight to get a desirable territory, they build intricate nests, or they may even construct an alluring trysting site, an elaborate boudoir used solely for the purposes of copulation.

Often singled out for their incredible collection of avian memorabilia, the male bower birds—relatively plain, crowlike birds of New Zealand—build elaborate bowers and spend months dancing within their decorated ballrooms in hopes of attracting a female for a few moments of copulation. The bowers are sights to behold: pavilions of twigs and grasses, festooned with flowers, fruit, bleached bones, snail shells, seeds, colored glass, and any other bright or colorful objects. The spotted bower bird is especially fond of bright and shiny objects and will even enter houses to pilfer enticing items, such as earrings, nails, screws, coins and bullets. The male of this species has a rather long and lonely vigil beside his elegantly arranged bower. He may have to display for weeks or even months before he finds a female willing to mate with him. His displays to any interested females are marvels of frenzied enthusiasm: He contorts himself and leaps about his bower. He even attacks his jealously guarded horde of baubles and tosses them about with his beak. Copulation, if it occurs, takes place outside his bower. Nevertheless he continues his vigorous display for a while before straightening up the disarray. J. C. Welty of Beloit College comments, in *The Life of Birds*, on the male's long bower-side vigil: "It seems that the long maintenance of sexual activity and display by the male — sometimes four to five months — is an adaptation to ensure the readiness of the male for the sudden, brief, and unpredictable period when the female becomes sexually active."

Thus, the lengthy display of the male does not induce egg maturation, which seems to depend on unknown factors — on the vagaries of nature and not on any such predictable calendar variables as lengthening days or phases of the moon; ergo, the male's long-term preparations are a convenience. He is ready as a sperm resource anytime she is ready for him. Sperm development has been primed by his bower building, and the female need do nothing to sexually stimulate him.

Among some mammals, if the male isn't ready with viable sperm when she comes into estrus, the female has a few tricks up her sleeve to guarantee that sperm will be available when her egg is ready to be fertilized.

Normally, both female and male rhesus monkeys are sexually ready during the regular breeding season (from late fall to spring), the males have volumes of sperm produced by their enlarged testes, and each female has brief flings with several males around each of her estrus periods. During the summer, both sexes go into a hiatus; testes shrivel, sperm production is phased out, and the ovulatory cycle ceases. But John Vandenbergh and Lee Drickamer, reproductive biologists of the North Carolina Department of Mental Health, have thrown a monkey wrench into the accepted explanation that the physical environment alone controls the rise and fall of sexual readiness. They experimentally induced sexual heat among females during their normal period of sexual inactivity (summer) by removing their dormant ovaries and revving them up sexually with estrogen to bring them into estrus. Then they released them to socialize with the males, whose summertime testes had become completely nonfunctional. The presence of these courting, soliciting, presenting females had an astounding effect on the males. Even though the temperature and day length gave nonbreeding cues, the males geared up for sperm production, enlarging their testes and starting immature sperm on their road to maturation. Thus, if the female rhesus found herself in a situation where the male had no ready sperm, she had the ability to bring him up to her sexual standards. She not only used the male as a sperm resource; she made him into one as well—a true fail-safe system. We don't know what happens most of the time under natural conditions, and experiments such as Vanderbergh and Drickamer's only serve to point out that there are many subtleties in the social-sexual system that we have yet to reveal. What such discoveries suggest is that the male may need a little help from the female to prod his sexual development and that just as the sage-grouse hen needs to be revved up by the sight of hundreds of courting males, so males of other species may need to be stimulated by the overtures and solicitations of amorous females.

One need not accept the stereotype of male domination over sexual matters, because, as we shall see, females in many species are in control of the quantity and quality of the male's sexual experiences. Even the most classic types—dominant, brawny, fierce—do not escape the power wielded by the female. A classic stereotype of the domineering male is the northern fur seal, who does not really dominate. His autocracy is limited to bullying other males about territory.

In many species of pinnipeds (seals and walruses), the male, who is much larger than the female, is traditionally depicted as a harem master, a collector of many females, a controller of their destiny and a feudal territory holder who can fulfill his sexual desires with any female at any time. The mystique surrounding male prowess originated from accounts of observers who saw only the male's aggressiveness to other males, his constant enforcement of his territorial rights and his apparent herding of females, keeping them within his home range. Yet if these early observers had watched the females, they would have discovered that the apparently passive sex was not dominated by the male but, on the contrary, was rather free-wheeling and controlled the male's reproductive success by being the active courter. Furthermore, herding resulted from female sociability, rather than from a fear of male retribution if she strayed.

Females congregate whether or not a male is present, because they are gregarious and find comfort in each other's company when giving birth and nursing. The females move in and out of a male's territory with impunity, and the harem "master" is hard-pressed to control his charges. By and large, the females ignore the male, hardly glancing in his direction even when he is locked in a biting duel with another male. The female covets him only when she is in estrus, right after she gives birth. Then she courts him with forthright gestures. Waddling over to the bull, she rubs her nose on his and nuzzles his neck. He responds with a great roar and then sniffs her vaginal region. His roar and sniff indicate his interest, and she continues her courtship display by arching her back and spreading her hind flippers, an unmistakable invitation to mount. The whole sexual episode takes about an hour, after which the female returns to her newborn pup and to the company of other females, once again ignoring the male.

In a cousin species, the California sea lion, the females seem, if possible, even more indifferent to the male's brawn, aggressiveness, and territorial skirmishes with other males. These females do not follow the males, attend to them, or even stay within a fixed territory. Instead, they move up and down the beach en masse as the tide flows and ebbs; unlike human sunbathers, who prefer high, dry sand, the female sea lion likes it wet. The males follow them, constantly adjusting their territorial positions to fit those of the females. A female in estrus announces it when her vulva becomes conspicuously swollen and pink, but the laid-back male doesn't even notice this blatant sexual signal. And the estrous female, pressed to get her ripe egg fertilized, seduces the male. She invites his attention as she languorously lies prone in front of him, pressing her body against his, writhing on the ground and over his back. A responsive male begins to sniff her genital region, and she invites him to mount by arching her back and spreading her hind flippers just like her cousin.

The male mounts several times until she is satisfied and then she breaks off the tryst unromantically, by biting him and pulling away. If the male is a hopeless dud, uninterested in her advances, she waddles off to solicit another, luring him with the same sexual displays. One mating is enough, and the female returns to nursing and tending her pup.

The all-powerful harem master has become a stereotype for male dominance. In pinnipeds, the image prevails; the male seal is dominant as beach master and feudal lord, tightly holding onto his harem of passive, submissive females. Males dominate males. Males fight and take out their aggressions on males. Dominance, which generally results from successful aggression, increases the odds that the winning male will be chosen by a female. Much of the myth surrounding mating—that is, males as the aggressive initiators and females as the passive recipients—stems from observations of male-to-male encounters. But it is the female pinniped who controls mating; it begins when she is ready and ends when she is satisfied.

Using the male as a sperm resource is not restricted to our mammalian cousins. Female snakes are no slouches either when it comes to getting males to service them during the brief time that their eggs are ready for fertilization. They advertise their sexual readiness by secreting a pheromone from the skin. A female adder slithering along the ground advertises her sexiness by leaving a trail of ophidian perfume, which a male can detect and follow to a sexual assignation. When a male adder, active in his search for females, crosses the trail of a sexually ripe female, he stops, turns his head left and right, flicks his long tongue to taste the trail, and soon homes in on her direction. Continuously flicking his forked tongue, he follows the route that she has laid down. When he finds the receptive female, he continues to flick his tongue but now along her sides and back, gently grazing her scaly skin as she whips her tail very fast. The male nods, flicks his tongue rapidly, and quivers in his excitement. The female may move away from him slightly to see if he'll follow, thereby testing his sexual intentions. He reaffirms his desires by pursuing her. She permits the male to enfold her and, undulating together, they line up, head to head and vent to vent. The male then inserts one of his two penises into her cloaca and, in mutual ecstasy, they wave their tails slowly about. During copulation, the female may become aware of her vulnerability, so she slides into a protective underbrush, pulling the male with her, for he is attached to her by his penis. Mating is brief. The female moves off to her single life and lays her eggs, her ophidian odor turned off until she is ready for the next sexual encounter.

Females of a tropical nocturnal spider known only by its scientific name of *Cupiennius salei* also leave pheromone trails to entice wandering males. A female coats banana leaves with her distinctive odor, and a male who happens upon the scene immediately begins a courtship dance vibrating distinctively with the harmonies of lovesong. The female, on another leaf, responds with her own vibratory signals, and her vibrations guide the male to her.

An observer of either snake or spider, unaware of pheromones secreted by the female, upon seeing the male pursue the female, might assume that the male initiated courtship. However, sophisticated techniques of odor analysis reveal that the female initiates courtship by laying down her seductive scent, which the male is impelled to follow. Male adders and male spiders, like so many other species, are strictly sperm resources, a mechanism to get ripe eggs fertilized, offering nothing more from the female's viewpoint.

Fathers Who Get Custody

But not all females contract to mate briefly and solely for fertilization. Some put the male to further use. For example, the female stickleback fish employs her mate to guard her eggs as well as fertilize them.

The courtship of the three-spined stickleback, a small freshwater fish, is a classic in the annals of ethology, the study of animal behavior. Typically, it is described from the male viewpoint. During the breeding season, the male stakes out a suitable territory, which he defends vigorously against all intruders and on which he constructs, out of plant fibers and sand, an elaborately woven tunnel nest. He then parades in front of this boudoir, displaying his bright red belly to any passing sticklebacks. Should the passerby be an eggless female or another male, the resident chases them off his property, but if a gravid female laden with ripe eggs swings into view, he zigzags crazily toward and away from her. If she finds these mad zigzagging dashes irresistibly alluring, she points her head up and shows him her swollen belly filled with eggs awaiting fertilization. The sight of her swollen belly apparently satisfies him that the female intends to make a deposit in his nest, and he leads her to the nest, pointing his snout toward the entrance. She wriggles into the tunnel nest, lays her eggs, and exits out the other end. He follows her, fertilizing the eggs. With no love lost, reverting to his suspicious ways, he drives away his erstwhile mate, chasing her off his territory. This one-sided story upholds the stereotypical pattern of male-initiated courtship. But further study showed that under some circumstances the female makes the first pass.

The sticklebacks of Wapato Lake, in Washington state, have been quite closely scrutinized by scientists eager to understand them. Following them through an entire breeding season, observers discovered that the "typical" male-initiated courtship occurs only early in the breeding season; later on, things change. Early in the breeding season, a female can be choosy about the male she allows to father and protect her offspring, because every male has an empty nest, and hence every male, eager to mate and collect eggs in his nest, will dance his zigzag best before every swollen-bellied female. She can swim around and choose the right combination of male, territory, and nest that approaches her piscine ideal. Later on in the season, though, her options diminish. Males stop actively courting when their nests are full of eggs — and furthermore, in their newly protective paternal mood, they view all approaching fish, gravid female or not, in the same light: as "infantivores" to be routed as quickly as possible. Now the female must vigorously seduce the reluctant male. When the female initiates courtship, she pushes her rounded belly time and again into the snout of the male, who usually responds not by pointing to the entrance of the nest but by chasing her away from the door. Late in the breeding season in Wapato Lake, any male with a nest is constantly courted by several females who — undaunted by his obvious, often vicious, attacks — are desperate for a chance to lay a clutch of eggs.

The male's unwillingness to fertilize eggs has not to do with his lack of sperm but with his lack of space to house the eggs. In effect, females have created their own dilemma by opting for male guardianship. A male, unable to fit more than a thousand eggs into his protecting nest, can't accommodate all willing females and probably can't protect a nest larger than the one he has. So some females must accept rejection in exchange for his aggressive territoriality that gives eggs a bit of protection from marauding mouths.

Another fish who uses the male as more than a sperm resource is the pupfish, a genus of chubby little fish whose main claim to fame is an incredible tenacity in clinging to life under the most difficult conditions. Indeed, these fish positively thrive in such places as tiny natural wells and warm springs in the midst of the hostile and decidedly unaquatic deserts of Death Valley.

Researchers peering into the details of their reproductive habits rapidly discovered that, like the female stickleback, the female pupfish exploits the aggressive, territorial nature of her mate to protect her eggs. The male pupfish doesn't have the paternal instincts of his stickleback counterpart, for he neither builds a nest nor tends to the offspring; his paternal duty is the vigorous defense of his territory and, incidentally, any eggs included therein.

Pupfish courtship is brief and to the point. Males patrol a parcel of real estate in shallow water containing a prime site for egg laying: a nice bit of rock, perhaps, or a clump of feathery vegetation. He colors his back a distinctive metallic blue and swims incessantly from one territorial border to the next, ever alert for trespassers.

Fertile female pupfish eye the males from the safety of the surface water and choose a likely-looking territory owned by a likely-looking male, one who is, for reasons unknown to biologists, more or less the same size she is. Each female then enters her chosen male's turf. He, ever vigilant, begins his courting pursuit. And she, accepting his amorous advances, descends with him in a tight spiral to the bottom, where they spawn.

The female pupfish is among the most thoroughly promiscuous of creatures, spawning with many males and releasing only one or a few eggs in each male's territory. Her promiscuity, by evolutionary design, serves her well. Since the eggs are looked after only minimally, the female smartly lays her eggs in many of these quasihavens—in that way, chances are good that some will make it to adulthood.

Another aquatic creature, an insect, uses the back of the male with whom she copulates as a perambulator. Four inches long—giants among insects—the female waterbug lays her eggs on the male's back, using it as a nesting site. Typically, a female with freshly ovulated eggs approaches the male and engages in preliminary love play (called "sparring" by some observers), which probably tests the male's interest. She knows that he is ready to copulate when he performs a really spectacular act: He pumps himself up and down, strangely reminiscent of human copulatory thrusts. Far exceeding human stamina, however, the male waterbug may pump three hundred times a minute for two minutes. Evidently, pumping is the female cue that he is ready to fertilize her eggs. She clambers onto his back, and soon copulation ensues. The male gets himself in position to guide the eggs, using special organs to feel them, onto his back as the carefully oriented female extrudes them. When it is all over, the female rests and grooms, and the male is left with his new charges. Revived, the female may copulate again with that male or she may find another pumping male who is more attractive — another back for her young.

However, these females also encounter the same problems as the sticklebacks. Once a male has a load of eggs on his broad, flat back, he may refuse to copulate, and the female must hunt for another willing consort and egg carrier.

From these examples, we see that for some species' females, the males serve them, beyond sperm contribution, as guardians and protectors of the developing embryos. Freed from child care, those females may cavort with other males, who, in turn, will watch over eggs if the females deign to leave them in the males' nests or territories, or on their backs.

Bargaining for Sex

Among other species, the way to the female's heart is through her stomach. Courtship brings with it sperm and lots of protein. Butterflies have very limited diets, feeding only on the sugary nectars of flowers, almost pure carbohydrate. Yet butterfly eggs, like other eggs, require lots of protein if they are to become embryos. Thus, at egg-making time, the female should, like the blood-sucking female mosquito, arrange for her diet to include protein. Can you imagine a blood-sucking butterfly?

The beautiful, much-admired butterflies, such as the orange-and-black monarch and black-and-yellow-striped zebra, have not resorted to such heinous behavior. To make her eggs, she calls upon two sources of protein — one from her own stores, left over from her caterpillar days, and the other from the stores of the male with whom she mates. The male willingly gives his stores to her under the most pleasant of circumstances, when, at mating time, he transfers his sperm package containing not only the wiggly masses of DNA-laden gametes, but also quantities of nurturant protein. The female makes good use of the contents. The sperm plunge headlong into the waiting mature eggs, and the proteins head for the eggs that will soon develop. These male protein contributions have been experimentally revealed by radioactive tracers, which were injected into the male and later discovered in the eggs and tissues of his mate. The monarch and zebra do not restrict their amours to just one male. They collect lots of protein by being promiscuous. Many males means protein for many eggs; in the monarch and zebra, successful mothering comes from multiple matings.

Using a sperm packet for nutrition as well as fertilization is not as deadly as the game played by the praying mantis, an insect notorious for its ruthless mating habits. The female of this species eats her mate while he is inseminating her, not because she is dissatisfied with his copulatory technique but because she needs the extra protein to enrich her eggs. Other insects, such as the desert katydid female, are often faced with water shortages. Her mate presents her with sperm packets that are waterlogged, thus alleviating a serious problem: where to find water to put into her eggs in the middle of an arid habitat.

Many choosy insect females, who must often get more than sperm from their males, have developed very sophisticated techniques, qualities that remind us so disturbingly of human courtship that we struggle to keep from anthropomorphizing them. Such patent analogies are demonstrated quite dramatically by a master of the art of gift-giving courtship, the black-tipped hanging fly, who has made it to the front pages of popular scientific magazines through the promotion of its scientific scrutinizer, Randy Thornhill, a biologist at the University of New Mexico.

The elaborate courtship begins when a hanging-fly male, fortunate enough to capture a fat housefly, alights on a branch, dangling there and exuding sexual pheromones to lure a female to him and his booty. A sexually receptive female, unable to resist his perfume, flies over to inspect his catch. He clings to his booty as she touches and tastes it. If the offering is big enough, she proceeds to suck out the juices, and while she dines on the fat housefly, the male copulates with her, sometimes for as long as forty minutes. In the end, she lays about three eggs.

The hanging-fly female has evolved an effective reproductive system. She has reduced the risks of hunting—such as getting entangled in a spider's web—and the energy drain of an hours-long search for a delicious daddy longlegs. Yet she manages to get as much protein as she needs to produce eggs by simply making the male bring home the bacon. And if the food offering is too small, she rejects it and the male, refusing to copulate.

According to Thornhill, an insect offering with a surface area smaller than sixteen square millimeters is a complete reject. Somehow, in her dull insect brain, she "knows" that an insect of that size or less is too small to make mating worthwhile, because it will not make a significant nurturing contribution to her eggs, even though the male's sperm may be adequate.

Strictly speaking, the "gift-giving" trait is misnamed. The prey insect is available only during copulation. Afterward, males scuffle with the female to keep the gift, to lure a new female to dinner. Humans, more subtle in their courtship and gift-giving tendencies, oddly parallel the behavior of the hanging fly. How many times does a courting man give a woman gifts — gold bangles, a fur coat, or a television set — with the expectation of copulation in return? And how many times have women had to deal with men who, like the hanging flies, want the gift back after the copulation is over, so he may lure yet another female?

Choosing a male on the basis of the size of his gift is rare, but choice based on other criteria is more widespread. Bigness, loudness, territorial quality, or aggressiveness may make one male more desirable than another. For example, among fish, female mottled sculpins lay their eggs on the ceilings of tiny, rocky caves that are aggressively guarded by the male. Bigger males are more adept at chasing off potential egg eaters, so the female who has ready eggs cruises from one male's territory to another, literally sizing up the available males and choosing the largest among them as her mate and the protector of her eggs. Science is baffled by how she measure the male, especially since the female must remember the sizes of males seen earlier in order to compare them.

Male European wrens build nests to entice a female, and they vigilantly patrol the borders of the territory in which she incubates the eggs, as well as find food for the nestlings. Apparently, a female surreptitiously makes the rounds of the male's territories, comparing the relative merits of nests and acreage, and upon making her choice of a desirable house and garden, she presents herself to the nest builder, to be courted. Here, too, her size-comparing techniques have escaped the eyes of prying observers.

Male desirability is sometimes evaluated by the tune he warbles. Many monogamous male songbirds who sing so melodically in the spring, use their different songs to let the various females know which ones are the proper species to mate with. Many species of sparrows look so similar that at a distance they can be mistaken for one another. But their different songs make them quite distinguishable. A female song sparrow needn't waste any time checking out a courting swamp sparrow, since she is attracted only to song-sparrow melodies. Oddly, although she opts for a singing mate, her preference doesn't seem to be dictated by her genes. Rather, she prefers the songs, which she heard as a nestling and fledgling, these are the ones that titillate her in adulthood. Carrying preference to an extreme, white-crowned sparrows distinguish between various subtle dialects within their own species and choose the males whose song most closely resembles the dialect heard in early life. These females truly want a male just like the male that married dear old Mom.

Among frogs and toads, croaks have sexual allure, too; females are attracted to the males with the loudest voices (who are usually the largest as well). How loud croaking and bigness serve the female's needs is moot, for neither parent guards the eggs. In one tree-frog species, the loudest callers are those who sit on certain branches that give their croaks particularly resonant characteristics. These well-amplified croaks attract the greatest number of females.

In another species, the lemon tetra, a common aquarium fish, the female scrutinizes the male and looks for "new" males instead of "used" ones. A male who ejaculates often with many females within his display area suffers a decline in sperm count, and his ability to fertilize eggs drops off after several ejaculations. A female prefers to spawn with a male who has not recently mated, but how she knows remains a mystery. Maybe his vigor sags and his ardor cools after several matings.

Females of the majority of species do not devote very much time to preparation for courtship. In most instances, it is a brief fling, a momentary encounter, when compared to the time devoted by males, who may spend days or weeks or even months preparing nests, courting attire, and courtly manners. In the past, this wide disparity in preparation was interpreted to mean that the male was the instigator of courtship. But, when seen from the eyes of the female, it actually reveals something quite different. The fine hand of evolution has created an elegant strategy designed to serve the female. It has arranged to have the male ready for the convenience of the briefly fertile female, the brevity of her fertility dictated by the singularly short life span of the mature egg. The immensely successful strategy of making the male a continual sperm resource during the breeding season is attested to by its prevalence in thousands of unrelated species, representing vertebrates and invertebrates: bull frogs, red kangaroos, dragonflies, fiddler crabs, cleaner wrasses, birds of paradise, weaver birds, Anolis lizards, redwing blackbirds, and crickets — to list but a few.

One common characteristic of species in which the female courts the male briefly (i.e., engages in short courtship) is that the fate of the eggs is of no concern to either parent, or, if there is concern, only one sex bears the burden of parenting. Among many of the twenty thousand or so fish species, the female may abandon her eggs to their planktonic fate, put the eggs in a nest guarded exclusively by a male, or carry them in her mouth until they are a few weeks old. In all instances, care after fertilization is provided in a one-parent household. Short-courtship practitioners among birds and mammals follow the same pattern, but in most cases, the female takes responsibility for protecting and nurturing her immature young.

Short courtship works in these species because the female spends no time with the male after she mates. Once a male meets her basic requirements, she copulates and expects nothing further. But when the female is going to share parenting with the male, she scrutinizes him for much longer periods before she finally mates with him, testing his eligibility in such talents as nest building and food gathering. In contrast to the short courtship that characterizes species without shared parenting, long-courting species go through extended engagements for days or even weeks.

Long-Term Pair Bonding

In the process of this long exchange, the female and male establish what is known as a pair bond. Most people would call this mutual interest — the constant desire to remain together, this awareness of each other's state — love. When a pair bond is forming among birds, for example, the adults are making a truly serious commitment to one another as they exchange behavioral rituals, one behavior following another, organized and sequenced by the species' genetic code. The length and complexity of the rituals permit the female to assess the male's qualities; in virtually all bird rituals, for example, there is symbolic gathering of food.

Courtship in the Western grebe is a classic example of the complexity and involvement of pair-bonding rituals. The grebe, a large black-and-white diving bird with an elongated neck and slender harpoonlike yellow bill, swims in the lakes of the northern Great Plains. Living gregariously together, the females and males are plumed identically, but their voices and utterances reveal their sex. During the spring breeding season, unmated females advertise their single state by uttering a distinctive "creet" call in a high-pitched voice. An unattached male interested in her courting overture answers with a slightly deeper "creet" call. After a few "creet" exchanges, they approach, stare at one another, and, if they are attracted, firmly remain side by side, ignoring the unmated grebes' attempts at disruption of their courting.

Heads low, crests raised, throats bulging, the pair point their bills at one another and emit a harsh gear-stripping noise. Stimulated by this cacophony of sounds, they dip their bills and vigorously splash the water to one side. They take turns splashing, in sequences and in harmony, adjusting the timing of their dips to fit each other's rhythms. After a few minutes of splashing about, they lunge forward and upward, running quickly across the surface of the water, side by side, like dancers doing the Peabody. Suddenly they stop and dive, head first, into the water. Upon emerging, they stretch their giraffelike necks as high as possible and resume their intense staring at one another. Satisfied with each other's looks, they get down to a more committed part of their relationship. Each grebe dives to the bottom, collects aquatic weeds, and surfaces with the weeds dangling from its bill. These very same weeds will become the building materials for their nest.

Weed gathering is a critical behavior, comparable to a woman accepting the proffered engagement ring. It's at weed-gathering time that the female may break off the engagement. For reasons that escape scientific observers, she often refuses to collect weeds, even though the male is gathering weeds with passion. Indeed, of all initiated courtships, only three out of one hundred females consent to carry out the next step — that of the weed dance — which precedes nest building. On the other hand, if the female desires the male as her mate, she indicates her acceptance by gathering lake weeds with him. United, the female and male then announce their long-term bond when they cross their weed-laden bills and stretch upward. In addition, they reaffirm the attachment by churning their webbed feet and dancing with crossed bills. Finally, with a toss of their heads, they discard the weeds, sit down again, and preen their feathers in unison, mirror images of one another. Clucking contentedly, they move off to search out a suitable site where they will build a weedy nursery and raise their offspring together.

The female Western grebe appears to be very choosy. She goes through the preliminary courtship rituals rather casually, but when she must make the commitment of picking up weeds, ninety-seven out of one hundred times she draws back. Obviously, she is in no rush, taking her time to survey the male's potential. Perhaps his way of gathering weeds indicates the pluses and minuses of his parental potential; only the grebe knows.

Long-term pair bonds among geese and swans are idealized by humans as perfect marriages, filled with fidelity and loving cooperation until death do them part. Unlike birds whose pair bonds last only through the breeding season, geese and swans mate for life, breeding as constant pairs year in and year out. Furthermore, if a mate dies, the surviving spouse, faithful to its memory, refuses to mate with a new partner — at least that year.

The Canada goose, a big waddling bird with black neck and white cheek patches, pairs during its second or third year of life. The gander, undiscriminating, greets all other birds by attacking them. He hisses, honks, chases, and even bites, not distinguishing between females and males. It falls upon the female not to attack in retaliation. If she is interested in pairing with him despite his bad manners, she expresses her attraction by remaining near him and following him. In essence, by following him, she is reducing his aggression, calming him and getting him accustomed to her face. She doesn't venture too close at first, staying safely behind so that he can see her out of the corner of his eye. Soon she becomes a familiar sight, and with familiarity comes male acceptance; he no longer attempts to drive her away. However, the female is not ready to show fidelity, for although she is apparently taken with this male, she can easily leave him if another male happens on the scene and, in turn, follows her. To the male of her choice, she signals her serious interest in making goslings by dipping her head underwater and tossing the water over her back — an unmistakable come-on signal of her willingness to pair, which she does not repeat until she is actually ready to copulate, many weeks later. In the interim, she and her new mate go house hunting, searching for a fine, food-filled territory in which to rear their young. The act of house hunting seems to fuse the pair into an even tighter bond, and, as though united in thought, they both attack and drive off intruding geese.

The male is particularly vigorous in attacking rival males. Evidently, defeating a rival is a feather in his cap, and he gives his mate news of his victories by snoring at her. Driving off other geese apparently is an essential ritual in establishing long-term pair bonds and in getting desirable territory.

At a suitable nest site, acquired either by moving into a vacant territory or by driving off the current occupants, the pair set up their nursery. Copulation awaits selection of the nest site, and the female cleverly does not produce eggs until she has a place to incubate them. Although the two have been together for weeks, the marriage is finally consummated in a matter of minutes, and copulation takes a few seconds more.

The chores of nest building fall upon the female. She pushes grasses and other nesting materials together while the male defends the site. The female incubates alone as the male stands guard. After the young hatch, the pair guard and rear the downy goslings together.

Long-term pair bonding is carried to an extreme in the mute swan, their bonds starting earlier than those of geese — while the swans are still adolescent — and lasting for a lifetime. Evidently it behooves the female to spend a year testing her mate's compatibility before she produces her first of as many as a dozen broods.

Mate testing, carried to the extremes of geese and swans, is rarely seen among mammalian species, because most females don't mate for life. Pairing is brief and to the point. Yet a few species love, honor, and obey each other for life. The beaver — a well-known model for human industriousness — is a monogamous creature. The female beaver pairs with a male of her choice while she is still prepubescent, at least a year before she will be sexually mature and ready to copulate. She embarks on her mate-hunting adventure by marking a parcel of waterfront with scent from her castorium gland. The scent marking invites beavers to her turf. Soon after they approach, the strangers discover that the alluring odor was merely a tease. She, instead of welcoming them, tussles and wrestles with her "invited" guests. Lars Wilsson reports that in captivity, the female will vanquish the male even if he is larger.

She chooses one of the vanquished as her mate-to-be; he then slinks around, avoiding combat with her, but staying as close as he can safely get. Within a few days, the female accepts him as a consort. No longer battling, they sleep curled up next to one another during the day. At night they often groom and chatter. Together, they hunt for and find a suitable den to renovate into adequate housing for their young, taking months to construct the complicated structures that distinguish a beaver pond. Upon their completion, the bride and groom cross the threshold, and the marriage is consummated. During estrus, the bride mates with her spouse in the water as they swim in unison, side by side, cementing their pair bond, which may last as long as twenty years.

One striking point that emerges from evaluating the very long courtships of geese, swans, and beavers is that these animals don't squander their reproductive years on setting up the pair bonds. Long courtship is for juveniles, and the pairs are firmly established by the time the female has her first eggs. Such behavior is reminiscent of some human cultures, in which marriages are arranged while the bride- and groom-to-be are still children. In the case of humans, however, society dictates the standards of compatibility, not the individuals who form the pair bond.

One of the biological mysteries about the beginning of long courtship among these birds and mammals harkens back to the story of the shiner perch female who mates when her sex-hormone level is low, a time when most females are generally disinterested in males. Goose and swan females also court and pair with males long before they are hormonally ready to produce their first mature eggs. The stimuli that prod them into an interest in the opposite sex are unknown. Oddly, the long courtship leading to pair bonding does not reflect a long foreplay prior to copulation. Once they are mature enough to copulate, the amount of time devoted to copulation is exceedingly brief and, within one year, may occupy only minutes. The duration of copulatory activities per se is no longer than that of the other females who pair with males only briefly and solely for the purpose of having their eggs fertilized. Obviously, the maintenance of long-term pair bonds is not dependent on copulation to reinforce the bond unless these birds have singularly fine long-term memories.

When they are juveniles, these pair bonders anticipate their future when parenting will be shared on an equal basis. The long-term relationship probably evolved not because mates are hard to find when you are looking for one but because both parents are needed to rear the young in these species.

Can we draw parallels to the human condition? What kinds of insights do these animal models yield for better comprehension of the strategies of the human female?

At first glance, it would seem that the human mating system has a lot in common with pair bonders. Like them, courtships tend to be long, and like them, reproduction is delayed until the courtship has been certified as completed, and the pair bond (also known as marriage) officially registered. The basis for animal long-term bonding is the need to be together when rearing offspring. Is such bonding relevant to the female human who has either few or no offspring? Why should a couple stay married, or even marry, if reproduction is not a goal? It's hard to imagine that any other species would spend so much time and energy in a mating pattern which makes no biological sense.

A female human who intends to remain childless could have her pick of mating patterns. She could have brief flings, a long-term relationship, brief flings while maintaining a long-term relationship with one male, and so on. Her biology and her ability to manipulate her reproductive potential frees her from the strictures of one style of pairing. The loss of estrus that characterizes woman means that she, unlike the other mammals, never senses the desperation to get her egg fertilized. Freed from the compulsive drives of estrus, from the raging hormones that propel other females into a whirlwind of mating, her biology does not dictate when and with whom to copulate. In addition there is certainly no biological reason why she must copulate with one partner. Yet perhaps there is a reason to partake of long-term relationships. Single humans crave the comfort of a partner and look to marriage as a solution to that ever-threatening problem — loneliness.

Although animals use their long-term courtships to test each other for compatibility and companionship, two qualities necessary for successful shared parenthood and for their life-style, some humans have substituted the means for the end. For them the goal of courtship and subsequent pair bonding is not so much concerned with raising children as with finding a partner with whom they can achieve compatibility and companionship. Sometimes children are incidental to the relationship. Nevertheless, for many couples, as for the pair bonders, reproduction is a crucial and significant aspect of their life together; and despite the burdens of child-bearing and child-rearing, women have babies.

7

THE TYRANT EMBRYO

MARRIED COUPLES DECIDE to have children for many reasons — among them, to see what it's like; to provide successors for a throne; to provide an heir for the family business; to recapture, through their children, the experiences of childhood; and to leave their name (and, incidentally, their genes) to posterity. Such reasons, culture-inspired perhaps, gloss over the true biological reason for having children — to assure the survival of the human species. Deep-seated biological urges cause us to perpetuate ourselves, making women accept the stress, the physiological upheavals, and the great discomfort of pregnancy.

Human pregnancy is relatively infrequent, however, when compared to the number of pregnancies experienced by the females of many other mammalian species, in which mature females are pregnant or nursing more often than not. Indeed, once the mammal reaches sexual maturity, most of her life is deeply concerned with mating, pregnancy, lactation, and teaching and training her babes.

But the females of most other species are not so intimately connected to the new generation. They have wide-ranging options, from total abandonment of the eggs in the sea, to warming and turning eggs in a nest, to brooding eggs in the mouth.

Mothers in many marine species are completely indifferent and oblivious to the fate of their fertilized eggs. The billions of microscopic eggs and embryos of thousands of species of molluscs, echinoderms, worms, and fishes floating in the zooplankton are testimonials to maternal abandonment. For the females of these species, their reproductive role ends when they spawn. Having no strategy for embryonic care, such species simply produce billions of eggs and never know how many meet an inglorious end, either as the dinner of some slightly larger creature or through starvation. These mothers make a small investment in their eggs, stocking them with a token amount of yolk. The newly hatched young, poorly nourished, are pressed to search for food (even smaller than themselves) soon after they escape from their transparent shells. However, as a reproductive strategy, producing billions and then abandoning them obviously works, otherwise our favorite oyster on the half shell and delectable milky clam chowder would be unknown to the human palate.

Guarding the Eggs

For other species, however, where females produce fewer eggs, lifestyle demands different strategies for egg care. Since egg production is relatively low, abandonment is too risky. To keep the species extant, such females must protect the vulnerable progeny and amply provide nutrients for each egg, to carry it from the two-cell stage to the point where the young can seek their own food; in some species, offspring dependency lasts for weeks. The techniques of safeguarding and nurturing are creative evolution at its best.

For the tree frog, protection and keeping the eggs moist are the female's main concern. She mates on a treetop branch where she extrudes her eggs. With a liquid secreted from her body, she whips up a foamy nest to surround the eggs. Within the stiffening foam, the young—laden with yolk; high, damp, and safe from predators—develop without benefit of further parental care. The first heavy rains dissolve the meringuelike nest, and the newly hatched tadpoles fall into the puddles below. In this case the female, like the briny creatures, has abandoned her eggs, too, but not before she has invested her efforts in converting her slimy secretions into an airy, moist nest that can nurture tadpoles, even in the arid environment of treetops.

Alligators eggs develop best in warm places; the female buries her shelled, yolk-filled eggs in a heat-generating, moist mound of fermenting compost that she has assembled. Although the eggs are hidden from sight, she patrols the area around the nest to drive off marauding racoons and other egg-loving small mammals who know how to dig out the buried eggs. Locked into her nest site and vulnerable for about forty days, her vigil ends at hatching time after which the young are essentially on their own. As they hatch, they call with peeping noises, and she helps them out of the nest, freeing them from the rotting vegetation. Before the eggs hatch, even when she is per-

THE TYRANT EMBRYO

sonally in danger, she is stubbornly bound to her nesting site and will not leave, attacking human intruders viciously and voraciously.

The alligator mother suffers no ill effects from her dedicated guard duty—unless her stubbornness allows her to be captured and turned into a pair of shoes. But other animals, such as the octopus, may give up their lives merely for standing guard over their babes. Laying strings of jelly-coated, yolky eggs in a selected nursery cave, she posts herself at its entrance to fend off intruders. Filling the entrance with her rotund body, she stays there, neither moving nor eating, ever watchful. By the time the young hatch, she has lost so much weight that she cannot recover from her octopus-style anorexia, and succumbs to death by starvation. Sadly, however, her dedicated self-destructive vigil does not assure the newly hatched polka-dotted octopuses of a secure future. Most are eaten as they leave the safety of their small nursery. Clearly, since octopuses are not extinct, a few must get away. However, if the mother did not guard, perhaps none would ever make it to the hatching stage.

Guarding requires extra parental investment, and if the female should leave the nesting site even briefly, for one reason or another, the embryos are singularly vulnerable to predation—as vulnerable as those produced by females who cast their spawn asunder. Guarding by mothers must be dedicated and unflappable, like the vigils of the alligator and octopus, who remain at the nest, curtailing or even putting aside their normal activities—including such fundamental ones as searching for food.

In a number of species, as we mentioned earlier, females beat the rap of guarding by using the male to protect the developing embryos (e.g., such fishes as the stickleback and the mottled sculpin, and species of polyandrous birds). Still, these females must depend on male dedication, and a male who has a roving eye or is engaged in territorial battle may forget about his responsibilities. When he returns to the nest, he may discover that a predator has taken advantage of his absence to remove his responsibilities.

One optimal strategy is for female and male together to oversee the welfare of their offspring and share guarding and incubation time. Sharing is the style of many fish species, such as the sergeant-major, the blue acara, and the jewel fish; and 90 percent of all bird species. Fish parental tasks include fanning to bring oxygen to the embryos, picking up debris that settles on developing eggs, and driving off intruders. Birds carry out their tasks by sitting down on the job and warming the eggs. They incubate, using their own body heat to raise the eggs' temperature and thereby speed up embryonic development. By sharing the responsibilities of guarding with the male, the female is not faced with the detriments of guarding alone, and she is not forced to remain in constant attention. She can exchange guard duty when it is time to feed herself. Even when she bears the major brunt of incubation, the male can bring food to her. Among ducks and geese, where only the female sits, the male remains nearby, giving additional protection to the nest when she leaves on daily jaunts to eat.

Females of such species have many options in caring for their eggs, which come from the strategy of egg laying, that of placing the eggs outside the body into a nest. The eggs, laid in some kind of nest — in a rocky crevice; on a smooth stone; in a mass of twigs, a treetop, or a sandy pit — are totally separate and can be guarded by mother, father, both, or neither. Some species, however, eschew an external nest and instead convert a portion of their bodies into a nesting site, where they place the fertilized, yolk-filled eggs. Naturally among these species, only one parent is committed to caring for the offspring.

Toting the Eggs

Practically any cavity, appendage, or surface of the body can be used to protect the eggs: the mouth, a cheek pouch, an abdominal pouch, the back, the area between the legs, or even an ovary.

Fishes that protectively tote their eggs around with them have accomplished the task in an amazing variety of ways. Some, such as the South American banjo catfish, stick eggs onto the surface of the female's belly; other species have special hooks on their heads that become festooned with clusters of sticky eggs. Several cichlids — those favorites of tropical-fish hobbyists — fill their gaping mouths with eggs. Although different species of cichlids vary as to which parent has the task of mouthbrooding — the mother, the father, or both — in all instances the pattern of care is similar. The responsible parent conspicuously "gargles" the mouthful of eggs several times a minute, churning them around in its mouth to clean them and aerate them in much the same way that the agitator of a washing machine cleans clothes. Ten days or so after the onset of mouthbrooding, the fully formed young, perhaps as many as forty, are unceremoniously spit out.

Probably the most notorious example of egg-toting in fishes is the bizarre "pregnancy" of the male seahorse. The brood pouch on the male's abdomen reminds us of a kangaroo's marsupium. The female seahorse places the fertilized eggs into the male's pouch, where the embryos, protected safely inside, even gain some fringe benefits: They may absorb oxygen and even nutrients through the spongy pouch lining.

Another form of piscine pregnancy is the retention of eggs inside the female's body. Many aquarium fish, such as guppies, swordtails, and platyfish, belong to a family called the livebearers, whose ovaries, the egg source, serve as their nest. In these species, sperm are transported up the female's reproductive tract and stashed in the ovary, waiting as long as several months for the opportunity to fertilize a maturing egg. Once fertilized in the ovarian nest, the developing embryos remain there, protected from the dangers of the outside world, only emerging when they are fully formed. They survive the embryo stage when fish are most vulnerable to being eaten.

Fish have the evolutionary option of toting their eggs around, either inside or outside their bodies, because their naked, permeable-shelled eggs are designed to survive in a watery milieu. It makes little difference whether that milieu is a gravelly nest, a sticky abdomen, a churning mouth, or a secretory ovary.

Females that live on land are not so freewheeling. Amphibians, for example, are caught in a particular bind. Their eggs, like those of fish, are thinly shelled and permeable to water, and therefore the egg nest must be in a watery environment. Most species simply mate in water and dump their yolky, jelly-encased eggs there. But some adult amphibians, especially salamanders and toads, live in a terrestrial niche often far from watery nurseries. Many of these species have developed strategies for carrying their eggs in a moist body nest—mouthbrooding like the African cichlid, or even brooding eggs in the stomach and regurgitating them at birthing time. One small toad of the African highlands broods eggs inside a specialized part of the oviduct during the dry season—putting them on a slow burner, as it were, for as long as several months. When the first rains of the wet season arrive, embryonic development suddenly speeds up, and very shortly afterward the young are born into a fresh watery milieu.

The midwife toads are the only amphibians who carry their eggs around in a nest on the outside of the body. In these bizarre species, the female sticks the eggs into special protective pouches on the back, where they are overgrown by the skin and develop, safely tucked away in their own little incubators. Some lizards and snakes also make nests from their reproductive tracts, offering protection to large, yolky, nutrient-rich eggs.

Birds never form internal nests, undoubtedly because of the constraints their weighty eggs would impose upon flight.

Parasitic Embryos

In all avian species, the mother enriches the ovum with enough yolk so that she need not supply additional nutrients, and the embryo, as it transforms itself, calls on its own well-stocked larder of protein and fat to build new tissues. If the female does not stock her eggs with all the needed amino acids, fats, vitamins, minerals, and other nutrients, she has to find another way of getting these essential body-building materials to her embryos. Females who have evolved toward making small eggs with little or no yolk had to opt for a new system of embryo provisioning. These females must supply the embryo with essential nutrients from their own bodies, thereby taxing their own systems.

In the group of the surf perches, for example, a score or fewer embryos float in an ovarian nutrient broth, absorbing so much maternally supplied food through their skins as to increase their weight twenty-fold or more. Among many sharks and rays, the mother and her few embryos attach to each other through a placentalike connection in the uterus, allowing the embryos to increase their weight three-hundred-fold or more. But such weight increases are trivial when compared to the most successful group of animals that produce microscopic eggs with virtually no yolk—the mammals.

Mammalian embryos get more from their mothers than any other species. Their efficiency should be studied by corporate kingpins who want to turn a small shoestring operation into a worldwide cartel—for that is the degree of growth and expansion carried out in

THE TYRANT EMBRYO

the embryos of many mammalian species. A mammalian embryo burrows into its mother's uterine tissue, makes a placenta, and takes its succor there, growing at her expense. It is very costly to the mother.

Like a parasite, this embryo feeds off the tissues of its host mother, controls its own chemistry, and manipulates the host's chemicals to suit its needs. Basically, the embryo is an invader — a genetically different foreign body — that in the normal course of events would be destroyed by the mother's immune system (just as, for example, a transplanted kidney would be rejected). To ward off antibody attack, the embryo puts up a line of defense, more impenetrable than anything the military has designed, which may jam the mother's own attack system, suppressing her normal immunologic response through mechanisms that have as yet to be understood.

In addition, the parasitic embryo forces the host's endocrine system to become subservient and to follow the mandates of the parasite, who tampers with the host's hormone levels. The human embryo, for example, mimics its mother's pituitary, secreting a hormone that prevents shedding of the endometrium (menstruation).

The fetus even determines its own birthday by manipulating its mother's hormones. Furthermore, it diverts as much as 12 percent of the mother's blood, bearing amino acids, carbohydrates, fats, minerals, vitamins, and oxygen to its own system, and spews its waste into the mother's bloodstream, burdening her with its excretion. Many women become anemic during pregnancy because the parasitic embryo gobbles up much iron to make its own blood.

To nurture the insatiable embryo, many pregnant mammals must eat 20 percent more food than usual, an easy matter for some of us, but not so simple for the mammals that must increase their grazing time or spend more hours on the hunt.

Clearly, the mother gets little out of pregnancy, and indeed, it is detrimental to her health.

Could mammals avoid the inconveniences and stresses of pregnancy and be more like birds? If the human female could fabricate a yolk-filled egg, completely stocked with a nine-month supply of nutrients, the egg would have to be the size of a watermelon in order to hold her developing fetus; an elephant egg would have to be the size of a steamer trunk; and a blue-whale egg would have to be the size of a compact car. Yet a few mammals, albeit bizarre, do lay eggs, circumventing the egg-size problem by meshing a birdlike strategy with a mammalian one. These unusual creatures are the monotremes: the duck-billed platypus and the spiny anteater of Australia and New Zealand. After fertilization, they retain their eggs for about one month in the uterus, filling them with nutrients, encasing the eggs in a shell, and then laying them into an external nest or a stomach pouch.

The fact that only two mammalian species in the world are egg layers tells us that evolution has determined that this strategy is not optimal for mammals, not even if the eggs are small and manageable. As an experiment in evolution, egg-laying mammals will soon be extinct. In contrast, mammals that establish placental connections are eminently successful, filling all ecological niches, exploiting the oceans, tunneling underground, living in ice floes, migrating over vast plains, and even flying in the air. This mammalian explosion came about because the female mammal tolerates a parasitic embryo, despite the physiological detriments of internal gestation. No matter where the female lives, no matter what species she belongs to, embryos all grow in basically the same kind of environment. The polar-bear female, searching for carrion on the ice floes of the Arctic, keeps her developing embryo at the same temperature and with the same food supply as does the tiny dik-dik searching for acacia leaves on the equatorial plains of Africa.

Indeed, among such diverse species, pregnancy is remarkably similar, alike in both marsupials and placental mammals. Although the length of pregnancy may vary from two weeks to twenty-two months, the process is the same. All mammals make a placenta, a remarkable disposable organ that endows these diverse species with a uniform ideal environment. The amazing placenta is a joint venture between embryonic and maternal tissue. It regulates the exchange of nutrients and waste products, and the developing embryo does not become a parasite until the placenta is formed.

The placenta functions as a homeostatic mechanism, maintaining an internal equilibrium for the embryo, even under conditions that drain the mother's reserves. The placenta gives life. Without it, the mammalian embryo would die of starvation in a matter of days, as well as being poisoned by its own waste products. At the same time unspecialized and specialized, it duplicates the function of several body organs. Operating independently of the maternal endocrine system, one of its most important functions is the maintenance of the hormonal secretion. Like the endocrine glands, it secretes a variety of hormones; like the small intestine, it absorbs nutrients; like the kidneys, it selectively filters out chemicals; like the lungs, it supplies oxygen and disposes of carbon dioxide.

When gestation ends, the placenta steps away from its critical involvement with embryo and mother, and permits the soon-to-be-born infant to decide the moment of birth. In one extensively studied mammal, the sheep, the embryo's pituitary produces a hormone that forces the mother, by secreting her own hormones, to initiate uterine contractions — commonly called labor.

For most mammals, the birth process seems relatively easy, but for humans, it is often a traumatic experience, especially the first time. Modern human involvement with the birth process has resulted in the flowering of birth centers, as well as how-to courses on giving birth, such as natural childbirth, prepared childbirth, the Lamaze method, the Bradley method, the Leboyer method, and the water-tank method. The options for place of birth range from a standard hospital, to alternative birth centers to the home. The blossoming of these methods reflects the anxieties felt by most women; birth is an experience to be feared.

But only the female human, with her enormous capacity for cerebration, anticipates birth with trepidation and imagines the pitfalls of infant deaths, her own death, and unbearable pain. Furthermore, as birthing time approaches, she clings more and more to her family, to associates and to other women in the same condition who serve as support systems.

How unlike other mammals, where birth is a simple matter. The mother does not need assistance; her offspring, often in multiples, are smaller in relation to her birth canal than human babies, and she drops each one during what we consider an easy labor. For example, the common marmoset monkey may need less than a half hour to expel the first offspring and only five more minutes for the second.

After birth, all mammal mothers vigorously lick the young, cleaning off amniotic fluid and birth membranes. And in virtually all species, whether carnivore or herbivore, she then eats the placenta. Apparently in a state of intense pleasure, she ignores the mews or cries of her newborn, who may be dangling from the umbilical cord as she chews the cord up. It's easy to understand a carnivore like a cat or dog consuming the rich, spongy, blood-filled tissue, but what about the herbivores, such as deer, rabbits, and even most primates? What is the placental appeal? How does the ruminant stomach, designed for digesting grass and leaves, digest this high-protein, meaty tissue? How does the elephant stomach, also accustomed to leaves, digest it? Nobody knows.

Giving Birth: A Solitary Experience

In many species, the elaborateness and extensiveness of prenatal preparation is inversely related to degree of maturity in the newborn. Females that bear altricial (immature and helpless) young must provide a safe haven for a considerable period of time as the young mature outside of her body, while females that bear precocial (mature and walking) young require little or no preparation. In both types of species, the female finds a secure, secluded area.

For example, the pregnant sheep, normally a herd dweller, separates herself from her cohorts as parturition (birth) becomes imminent. A female bighorn sheep seeks out the most inaccessible spot on the mountain and may stay there, isolated, for days without food and water while awaiting her newborn. The elk cow, not quite so extreme, seeks out a shielded retreat at the edge of the herd. The rhesus monkey moves away from her troop to the edge of the forest and picks a well-camouflaged hiding place so that she may give birth alone, away from the curious eyes and unwanted attentions of other troop members.

Mammals that don't have the option to move away from the group still try to isolate themselves from their species. Such mammals as the nest-building rat and the horse give birth during the normal periods of species inactivity. The rat, normally a nocturnal prowler, gives birth during the day, and the horse, normally a daytime grazer, gives birth during the night.

This kind of removal seems strange to us because under such conditions the parturient female is particularly vulnerable, cannot protect herself from predators, and should, by reason, seek to be in the center of the group and to use group members as protectors from potential danger. The standard explanation is that by withdrawing from the herd, she becomes a small dot on the landscape, less conspicuous to predators.

THE TYRANT EMBRYO

Recent studies, however, suggest that it is not predators that the female may be avoiding but her own fellow species members. For example, when the female marmoset is ready to give birth, she usually leaves her primate troop during the night, sometimes accompanied by her mate. While the troop monkeys sleep, she perches in a treetop and gives birth to twins. A scientist experimenting on captive marmosets eliminated the night part of the day-night cycle, keeping on the lights for twenty-four hours a day. Parturient females no longer had the cover of darkness to obscure the birthing—they were in full view, and they were in trouble. During one birth, the troop members flocked to the mother's side with intense curiosity about the newborns. Highly excited about the births, they proceeded to eat the placenta, and, probably by accident, the offspring as well. Such behavior makes us understand why evolution opted for hiding away at birth time.

In spite of the detriments generated by the trauma and drama of birth and the parasitic fetus, carrying the developing offspring within the body has its advantages for the mother. The female, whether she harbors her offspring within a uterus, pouch, mouth, or any other bodily carrier, is free to move, forage, sleep, and carry on all the activities associated with her species' life-style. Furthermore, if conditions become unfavorable (with the onslaught of such catastrophes as forest fires, flooding, or drought), the female mammal carrying her young within her can leave the area of danger to seek a safer haven. However, animals that place their offspring in a nest are, like birds, beholden to the nest's location and sometimes must abandon their young to save themselves.

The immovability of the eggs profoundly affects the adult's behavior, and if environmental conditions deteriorate, the adults can still opt to leave, but they must abandon their progeny, thereby losing their broods for that season.

Interestingly, female birds' physiology changes little during incubation and feeding of the nestlings, but her behavior is drastically changed, since she must first sit on the nest for about two weeks and later catch insects to feed her new chicks. The female mammal, on the other hand, suffers a huge change in physiology during pregnancy, yet her behavior seems untouched. But after parturition, her behavior changes, too. Her body produces milk, bringing with it a new behavior—the nursing of offspring.

8

THE MYTH OF MATERNALISM

FROM THE BODY'S physiological achievement—that of making milk from assorted chemicals—has come the myth of maternalism, the myth claiming that the mother is inherently better than the father at caring for offspring, that she is more nurturant, more patient, and more psychologically suited for bringing up the next generation. Lactation, the secretion of milk, has had more cultural ramifications than the secretion of any other bodily fluid.

A little milk goes a long way. An offspring must remain with its mother, since she is its only source of food, at least in the early days of life. A mammalian father, barren of milk and excluded from the nurturant process, can abandon the newborn with minimal disruption to his life, but a mother can't, because if she goes, the babe starves. Since mother and offspring join in the act of feeding, and since feeding during early life is frequent, what is more natural than a bond developing from this pleasurable sharing of time? Furthermore, as a consequence of the early sharing, what is more natural than the mother later being involved with all aspects of her offspring's maturation, learning, and protection?

Lactation has created the myth of maternalism, that only the mother is suited to taking care of the offspring. Indeed, the phenomenon of lactation may well be the root cause of the axiom "Anatomy is destiny." Yet despite the fact that lactation is peculiar only to mammals, the myth of maternalism pervades our perception of all animal species, and most people consider the females, or the mothers, as the primary caretakers. But in a vast number of nonlactating species, as we shall see, the male is the primary caretaker, or he shares care equally with the female. Furthermore, in some species, the female totally abandons her eggs to the care of another species.

Among 90 percent of bird species — among songbirds (Passerines), for example — female and male alike share care of offspring after hatching. As a team, they protect nestlings and fledglings, and provide them with such high-protein foods as insects and worms, which they catch on the wing or dig out of holes. A few other species, such as pigeons, parody the mammalian style by secreting a fluid from their crop, called pigeon milk, when they gaze upon their fluffy new squabs. Both mother and father regurgitate milk, an androgynous achievement which might be worthy of emulation by humans if they could find the physiological key. Behavior such as this has been long watched and admired. What human has not succumbed to the seeming tenderness of a pair of finches flitting actively to a nest to feed luscious bugs to their infant birds? Mother and father — hunting together, perched on the rim of their nest, dropping tidbits into the gaping mouths of infant birds — reaffirm our idealized image of the nuclear family: mother, father, and children. Yet if that style of care is unsuitable for the species, other strategies dominate. Not all bird species pair up to look after their young. Some find that if two adults are good, then three or more are better.

For example, such diverse types as swifts, swallows, magpie geese, pied kingfisher, king penguins, bushtits, nuthatches, wrens, grosbeaks, tanagers, jays, woodpeckers, terns, murres, and cuckoos all form cooperatives for the feeding of their young. Frequently, the cooperatives consist of families whose young of the previous brood didn't leave home after fledgling. Remaining with their parents, though quite able to feed themselves, they help feed the new brood and, in some instances, are not very much older than the nestlings being fed. According to ornithologist Alexander Skutch, a month-old long-billed marsh wren "began to pick up worms and catch flying moths, some of which it gave to the other fledglings," who were younger. In another example, "five young Common Bluebirds of the first brood, all less than two months old, diligently attended four nestlings of the second brood." Were the mother and father to disappear, those young might be able to step into their shoes and successfully rear the nestlings.

The utility of nest helpers has been scientifically validated by Glen Woolfenden, an ornithologist at the University of South Florida. Jays are famous for their communal breeding, and in many species the assistance of the yearlings is critical to the reproductive success of

THE MYTH OF MATERNALISM

the parents, success being measured by numbers of offspring fledged. Studying the Florida scrub jay, Woolfenden found that the yearling jays collect food for the young, defend the territory, and protect the nest from marauders. If a mated pair alone could fledge two young, then with the assistance of their older offspring, they would fledge three.

The extensively studied California acorn woodpeckers have helpers at the nest as well. They excavate nest cavities, sit on the eggs, gather insects and acorns, and drive off intruders, including other groups of acorn woodpeckers. When the sex ratio is skewed and too many adult males abound, the excess feeds the nestlings. Indeed, careful measurements show that the males bring twice as many morsels to the nestling than do the females.

In the arctic tern, where males outnumber females, unmated males often associate with a nesting pair, feeding the chicks alongside the parents. And the pygmy nuthatch male, also unmated because of a shortage of females, helps mated pairs by constructing nests, feeding the female while she incubates, feeding nestlings, and cleaning the nest.

The myth of maternalism and of the special relationship between mother and offspring is further shattered by the lifestyle of cuckoo birds. Anis, a kind of cuckoo, lay eggs in communal nests. All the adults incubate eggs and feed the nestlings, not discriminating as to which offspring belong to a particular parent. The European cuckoo is a master exploiter, going even further than communal nesting birds by placing its eggs in the nest of other species. She abandons her eggs to the care of the nest-owners. The foster parents incubate the cuckoo egg and feed the hatchling as conscientiously as if it were their own, even though they doom their own hatchlings. Its habits less than endearing, a newly hatched cuckoo pushes the eggs of the adoptive parents out of the nest, and thus garners for itself all of the care and attention. Not being able to recognize its own young is clearly a detriment for the host species, but a strong asset for the cuckoo.

As among mammals and many other vertebrates, there are avian species in which a single parent looks after the young—sometimes the female, sometimes the male. Among wood ducks, mallards, sage grouse, and bower birds, for instance, only the female tends the young after hatching. But in such species as the American jacana (Jesus birds), the tinamous, the rhea, phalarope, and the killdeer, a female mates with a male, then deposits her eggs in his nest, after which he incubates and protects the young, rearing them alone.

Single parenting among male birds should not be considered all that remarkable. After all, male birds are just as capable as females of sitting on eggs, of building nests, of guarding, and of feeding young, as the vast number of two-parent bird species so eloquently proves. Yet this behavior is viewed with astonishment among scientists, who consider it and associated activities as totally "unmasculine" and label it "sex-role reversal." In *The Life of Birds*, J. C. Welty of Beloit College writes of the tinamous and jacanas, "Through some evolutionary quirk in these birds, the role of the sexes has become largely reversed. The females are larger, having the brighter nuptial plumage, do the courting and defend the territory while the docile, drab males build the nest, incubate the eggs and rear the young." This sums up very concisely the orthodox view of paternal care as an unusual male behavior. But the orthodox view goes beyond characterizing it merely as an uncommon strategy within the normal range of animal behavior. Instead, as we mentioned earlier, the assignment of the phrase *sex-role reversal* affirms our preconceptions that docileness and drabness are strictly in the domain of the female, that only the female has maternal qualities, and that there is something abnormal, indeed "quirky," about a male who opts to stay nestbound and incubate, instead of parading around in colorful plumage and doing battle with other males.

Paternal care of the young, so "unusual" as to elicit special treatment among scientists, is merely one end of a spectrum of avian parental care. The other end of the spectrum — in which only the female incubates — is not viewed as strange or bizarre, but perfectly natural because that is what females are "meant" to do. When males become "maternal," scientists find it fascinating and almost illogical — perhaps evolution has some arcane adaptation for protective, nurturant males? This train of thinking has led to the use of the term *sex-role reversal*, whose widespread acceptance simply confirms how extensively and pervasively the myth of maternalism shapes our view of animal behavior.

Because fish are at a sufficient evolutionary distance from us and therefore somewhat resistant to anthropomorphism, the concept of role reversal has not been widely applied to them, but among cichlids, cottids, and gasterosteids, the father guards alone. Since egg-laying fishes don't face any nurturant constraints, it matters not which parent is involved with fish-fry rearing. The female fish is not predisposed toward maternalism any more than the bird.

Why should a maternalistic tendency be etched into the brains of females who lay eggs? Egg laying frees the female from further involvement; caretaking can be carried out by anyone. The options are numerous. Thus, among birds, fishes, and other animals, the myth of maternalism should not apply, and although the female provides the eggs, her biological "nature" does not necessarily predispose her to becoming the primary caretaker of her young.

Binding Mother and Infant

In contrast to the caretaking options of egg layers, female mammals labor under the constraints of lactation and are tied to their offspring. Presumably, since evolution has closed out alternative caretakers for the rearing of the young, maternal behavior is etched in the maternal genes and in the biochemical changes that parallel the act of giving birthing, which leads to the establishment of a behavioral connection: the mother-infant bond. This bond has been a major factor in creating and maintaining belief in the myth.

The behavior of such hoofed animals as sheep, goats, and cows has also served to reinforce the argument of biologically programmed mother love. At birthing time, these animals leave the herd and give birth in a secluded, protected area. Within a few hours, the mother recognizes the distinctive odors of her infant and will permit only that individual to nurse, unceremoniously butting away any alien young introduced by experimenters. Although the lamb, kid, or calf stands immediately and is perfectly capable of walking and suckling, its mother cloisters it for several days until, it is thought, a mother-infant bond is tightly woven. Mother and infant are intimately interconnected and responsive to each other through chemical cues, such as each other's smells. The mother is indispensable to the survival of her offspring. Without her, the babe dies. The one-to-one attachment lasts a long time, through weaning; in some animals, such as the red deer, it may last for as long as a year.

Superficially, similar mechanisms of bonding, where the mothers form deep attachments to their offspring, operate among other mammalian species. Overall, the mother-infant bond is considered to be a female strategy to assure both the survival of her offspring and the maintenance of her position as the only caretaker.

Nowhere is the myth of maternalism so pervasive as in our own species. Women the world over are made the custodians of the next generation; every society designates "Mom" as the primary caregiver. The responsibilities of child-rearing almost universally fall on the shoulders of the individual who did the child-bearing: the mother.

The close connection between the biological role of the mother as child-bearer and wet nurse and her cultural role as teacher and caregiver has led most people to conclude that the propensity of women to be the primary caretakers of children is innately psychological. Whatever that innate propensity is, they would argue, it can't depend solely on lactation, particularly since technology (i.e. packaged infant formulas) has liberated individual humans from the task of lactation. Infant formulas are popular substitutes for, or supplements to the breast. The maternal papa who, like the phalaropes, stays home to care for the kiddies while his wife earns the money is a rarity, a freak, in our society, and the wife and husband are seen as showing sex-role reversal. Movies such as *Kramer vs. Kramer* and *The World According to Garp* demonstrate the view of our society toward men who do want to take care of the kids: No matter how good they will be, Mother would still be better. Men who, through necessity, must take care of their children, wail about the inability of males to show the requisite patience, understanding, and diplomacy that proper child-rearing demands, and about their constitutional deficiency to deal with the multitude of details and the repetitious, tedious tasks involved in running a household. A widower with children recently wrote a letter to the editor of the *New York Times* in which he extended his own recognized inadequacies to his entire sex and exempted women from feeling the same boredom and frustration "because of their natural maternal instincts."

Many people sincerely believe that women are more suited psychologically to mothering than men are. They believe that, in addition to the anatomical badges of child-bearing—the swollen breasts for suckling and the wide pelvis that allows a full-term baby to pass through—there must be some fixed psychological tendency to mother, centered somewhere in that nebulous region between brain and endocrine gland. But Eleanor Maccoby and Carol Jacklin, psychologists at Stanford University, in their monumental book, *The Psychology of Sex Differences* note that "there can be no doubt that women throughout human history are perceived as the more nurturant sex, and are far more likely than men to perform the tasks that involve intimate care-taking of the young." And yet, they conclude, scientific evidence is so sketchy that "it is not possible to say whether adult women are more disposed to behave maternally than men are to behave paternally."

Yet the myth lives on, perpetuated at least partly by the obvious division of labor in child-rearing. After all, women are the ones who, if they work in the fields, carry their kids with them or, if they work in the office, are more likely to stay home when their children are sick. If there is no innate biological difference in maternalistic feelings—if there is no psychological predilection of one sex toward taking care of the children—then how can we explain why humans, with their diverse societies and life-styles, with the liberation from the physiology of lactation, universally designate the female of the species as the child-rearer?

We would like to present an original hypothesis, a suggestion as to how the unequal division of labor gets started and how it perpetuates the stereotype of the mother as the parent who is best suited psychologically to be primary caretaker. We propose that women mother not because of innate psychological propensity, not because of the physiological connection to lactation, but simply because women are the ones who are there when a baby is born.

In order to clarify this connection, let's get away from discussion of the human condition for a few pages and turn to maternal behavior in other species. If we look at mammalian maternal behavior beyond pregnancy and lactation, including all caretaking activities—such as retrieving, licking, teaching, playing with, protecting, and feeding the young foods other than milk—we soon discover that maternalism is not the exclusive tendency of the mother.

For example, from the moment of parturition, female rats normally behave most solicitously toward their offspring. They constantly look after the naked, helpless newborns, nursing them regularly, licking them, retrieving them when they wiggle off too far, and repairing the cozy nest they built in preparation for birth. Normally, male rats show none of these caretaking behaviors; on the contrary, when a male sees newborn rat pups, he will aggressively bite them and often eat them. In this species, paternal caretaking apparently does not exist. The gulf between the parental care shown by the two sexes is so wide as to appear unbridgeable.

In the early days of animal behavior research — that is, up until the early 1960s — it seemed as though the difference between female and male rats' response to their pups was an innate one ("instinctive," to use a buzz word of the time), a difference that was "wired" into the female and male brains early in their development. Parental care awaited the proper signal — in this case, the newborn pups themselves — to trigger one of two responses: "eat them" or "take care of them." Yet when placed under the scrutiny of careful scientific investigations, these assumptions, like so many others, were undermined and toppled by contrary evidence. It turns out that mammalian maternal caretaking, historically considered an instinct based on genetic programming, and unfolding in fixed patterns with predetermined schedules, is *not* programmed solely by the genes. Jay Rosenblatt, an animal behaviorist at Rutgers University, manipulated female and male rats in his search for clues about the biological basis of maternalism; he discovered that hormones helped this trait to develop but that the offspring were critical to the establishment of social bonding between parent and child.

He wrote, "Rather than existing preformed and on an endogenous basis alone, the maternal behavior cycle of the rat, like the reproductive cycles of the ring dove and canary, develops during the course of the cycle on the basis of the interaction of hormonal and external stimulation."

In other words, the transient hormones of parturition prime the female to nurse and exhibit other maternal care, but soon afterward the hormones dissipate and the rat pups become the stimulus for caretaking behavior. To make his point even more sharply, Rosenblatt did the unthinkable, at least for the 1960s: He designed an experiment to find out if the mere sight of newborn pups could evoke caretaking behavior in an animal that was not the mother and had not been hormonally primed by pregnancy and parturition.

He exposed newborn rat pups to virgin females and males. What would a virgin female, who certainly had the psychological caretaking potential but none of the hormones, do with offspring? The male was used as a control, and it was assumed that he, being singularly nonmaternal, would quickly devour the pups. Indeed, at first neither sex behaved maternally, and Rosenblatt had to replace each group of newborns daily with another batch the same age, until a stunning turnaround occurred. By the sixth day of exposure, both the virgin female and the male began to nurture the pups, both showing the typical caretaking behavior of a female who had just given birth. Both licked, retrieved, and even acted as if they were nursing the pups. The female's maternalism was not surprising; the male's was astonishing. Thus, given enough time, and by altering the environment, caretaking behavior was induced in the male. Evidently, the male has the latent talent to care for pups, but he is rarely given an opportunity. In the laboratory he is rarely present at their birth, and because of the female's reclusive nesting habits, he has little or no opportunity to see them soon afterward. Apparently, the tendency to take care of young is not programmed solely in female genes.

The rhesus monkey is another species whose males normally seem to lack paternal affection. Although not quite so antagonistic to infants as are male rats (wild rhesus males are occasionally seen carrying babies around with them), the best word to describe typical male behavior is indifference. However, primate researchers Gary Mitchell, William Redican, and Jody Gomber of the University of California, Davis, found that this indifference, or even mild antagonism, is not fixed into his behavior. When they exposed a rhesus male to an infant who needed care, they discovered that he could show "maternalistic" feelings as tender and solicitous as any shown by a female. Just as in male rats, it took time to wear down the initial resistance, to develop the care-giving behavior, but once that happened, these males made great mothers.

Evidence from other sources also points to the premise that caretaking in adults is induced by the offspring. The adult is made into a caretaker by the behavior of the young. Several elegant experiments carried out on rats and cichlid fishes emphasize the crucial control on parental care exerted by the offspring. Among rats, if the maturing young are regularly replaced by younger offspring (e.g., ten-day-old pups replaced by five-day-old pups), the mother will continue to care for them indefinitely in response to the age of the pups. Her maternal behavior is apparently cued in by pup age. Hence, an experimental mother rat could conceivably nurture fresh five-day-old pups for a year and never realize that they are not growing up.

Similarly, in experiments conducted by Evelyn Shaw, blue acaras, cichlid fish who share parenting, continue to guard offspring indefinitely if the young fish are regularly replaced by even younger ones. Normally, parental care ceases at fifteen to seventeen days after birth, but by constantly replacing the older fry with younger ones, parental care could be extended for months.

Scientists often hedge on conclusions drawn from laboratory experiments, because of the artificiality of the situation. But there is plenty of evidence from observations in the wild to show that the mammalian mother is not the sole caretaker among many species and that caretaking potential—"mothering"—is found among males, nonmother females, and even juveniles. In a number of social species, infant care is not at all the exclusive domain of the mother.

From Coatis to Samoan Natives—Sharing Maternity

According to field studies, among Old and New World monkeys, the caretaking role is so popular that observers have classified those individuals as "aunts." A newborn langur, for example, "magnetically attracts" other individuals in the troop and will spend about 50 percent of its time on the laps of "aunts," sucking at the breasts of lactators and nonlactators alike. According to Sarah Blaffer Hrdy, "It is the nulliparous females—those who have never had an infant of their own—who are the most eager and assiduous caretakers."

In her book, *Sex, Gender and Society,* Ann Oakley reports that communal breast-feeding is practiced by a number of preliterate peoples. She cites Margaret Mead's reports on Samoan children, who are frequently suckled by other women in the household. Among the Dakota, "Sisters share the breastfeeding of all their children between them," and among the Alor, "Every child has access to many breasts and the [natural biological] mothers are not more consistently available—because of other duties—than those of other women." Bororo women "consider themselves more or less equally available to all the unweaned children of the group," and Arrenta women "nurse each other's children."

Communal nursing is also the life-style of other mammals. Lion cubs, seemingly not satisfied with their mother's nipples as a source of milk, are notorious for soliciting milk from any adult female in the pride, and more than likely, if lactating, she will accommodate them. Coatis, tropical cousins of raccoons, live in groups in dense forests. Although females retreat from the group to a tree nest at birthing

THE MYTH OF MATERNALISM

time, when the litter of young rejoin the group at five weeks of age, they are nurtured by other females as well as their mother. Wild pigs, one of the few hoofed animals to have litters of helpless young, also exchange mothers and young, sometimes in a communal nest. In the all-female herds of elephants, calves are looked after for many years by all the adults and are free to suckle from any nearby cow when the pangs of hunger overtake them. Rodent young (e.g., rats) willingly suckle from any lactating female, and she in turn willingly lets the young nurse, though only if they are the right age. Adults don't seem to know their pups, and pups are not singularly attached to a particular adult. Even the most rigid bonds of maternalism that make mother goats butt away alien kids can be broken by extended contact. If an alien kid is placed next to a new mother but separated from her in such a way that it cannot be butted or kicked, she eventually accepts it as her own and allows it to suck from her teat. She apparently needs to become familiar with its smell and, once familiar, establishes a mother-infant bond.

Although males do not lactate, they nevertheless have important caretaking functions in many species. As we have seen, laboratory research shows that in the right experimentally rigged situation, normally nonpaternalistic male rats willingly engage in "maternal care" (licking, retrieving, posturing as if to give milk). Moreover, in the wild, the males of other rodent species are just as solicitous of young as females are—except for nursing, of course. The male beaver, for example, watches the birth of his young, shares in eating the placenta, and remains with the family from birth onward. He feeds, carries, grooms, and huddles with his infants. When his spouse goes on a foraging expedition, he baby-sits, protecting the young in the den. Along with the mother, he participates in teaching them how to build dams and search for food.

Wild-dog males share pup-rearing equally with females, including guarding and regurgitating food for them. In fact, the males are so capable of looking after their pups that after weaning the mother is completely expendable. Indeed, in one known case, five adult males raised nine pups when their mother died five weeks after their birth. Typically, one-third of the carnivores, all of which are social species (e.g., dogs, hyenas, foxes, martens, mongooses), the males socialize, feed, guard, groom, huddle, and baby-sit for the young.

Primates outdo even the carnivores when it comes to "maternalistic" males. In approximately 40 percent of these species, males care for the young. In some, males eagerly seek the companionship of young, sometimes trying to snatch an infant away from its mother for the apparent pleasure of carrying it about.

Among the marmoset and tamarin monkeys, it's hard to say who is the primary caretaker, the father or the mother. In the common marmoset, observed freely breeding in zoos, if the mother experiences difficulty giving birth, the father has been seen to assist with his hands, as well as licking and cleaning the newborn twins. Within a half hour after parturition, the father becomes attached to his young, and if he is carrying them, he becomes extremely possessive and scolds his relatives who want to touch them. The twins are returned to the mother for nursing. She increasingly rejects the young as the time of weaning approaches, but the father continues his solicitous and vigilant behavior. Among their relatives, the tamarins, the father carries the young all the time, and when the young cry out in fear, he — not the mother — rushes to their side to comfort them. Males of other New World monkeys (though not necessarily the father), such as capuchin, squirrel, howler, tit, and night monkeys also commonly carry young, relieving the mother of her charges much of the time, and offer the young solace in times of distress.

Old World monkeys, such as the great apes, our closest primate relatives, have a very different social structure. The male is generally excluded from the old-girl network. Hence, he rarely has an opportunity for contact with young. Among gorillas, the males seem to tolerate young, but they rarely, if ever, seek out their companionship. Yet, not totally indifferent in moments of infant distress, the male may come to the rescue.

Excluding the Father

One group among primates that especially interests us are the savannah dwellers, such as the baboon. Scientists think that our ancestors also lived in the same kind of treeless savannah habitat and organized themselves into similar group structures. Since an animal's behavior is intimately intertwined with its habitat, baboon social systems tempt us with analogies appropriate to our evolutionary ancestors.

Baboon females have the last word on virtually everything: the direction of travel, where to forage, who's going to be permitted to sit near whom, and who will come near a newborn. Males are just as eager to investigate an infant as the females are, but they are not welcomed as social equals. Males are excluded from the club until the babes grow up, and are forced to the troops' periphery, where they serve only as guardians and protectors. But when given the opportunity, a baboon male (not always the father) will look after juveniles quite assiduously. He will carry, touch, rescue, supervise play groups, and intercede in fights among young, defending them from danger and watching them in the absence of the mother. Furthermore, motherless infants are often adopted by a male, who provides the protection and care often associated with maternal behavior, showing an elaborate repertoire of parenting.

Some primate females establish old-girl networks and smoothly carry on their daily endeavors through a dominant-subordinate relationship that would make the female human reel. The baboons exemplify this pattern. Females of such other primate species as the chimpanzees also exclude males from the family circle. Thus, although maternalistic potential may exist, it remains forever dormant among such species.

Among langurs, infants are passed around, like a game of Musical Chairs, from female to female. Rarely is a male permitted to hold an infant; he is actively pushed away by the females. In the spectacled langur, "Within hours after birth babies are passed between adult females; and, while the male shows great interest, he is not allowed to handle the infant at this time," according to data cited by Gary Mitchell, primatologist at the University of California, Davis. In this species, however, after the babes become independent of the mother, the male is seen carrying, grooming, and playing with them (a social exchange initiated by the young, perhaps).

Many mammal groups have a social organization that precludes the participation of fathers. Females mate during estrus and then part company with the male. Hence, the fathers are never around at birth or during the postnatal development of their babes. They can't possibly be caretakers.

Of course, males who serve only as studs may be capable of caring for infants, it has been shown with the rat, but, short of more elaborate experiments, we will never know if such potential exists in their behavioral repertoire.

The potential for male "maternalism" may be universal, even if the opportunities are rare. As Gary Mitchell ironically puts it in his book *Behavioral Sex Differences in Non-Human Primates*, "Of course, there *are* those that believe the adult male human is quite capable of providing good infant care. The fact that even orangutan males show gentle infant care in zoos and laboratories suggests that the allegedly more intelligent human male may be able to do even better." And Maccoby and Jacklin wryly concur: "Even with little experience with infants, however, the human male may have more potential for nurturant reactions than he has been given credit for."

Men, in fact, have just as much potential as women to be good "mothers" — and considerably more chance of success at the job than have the males of other mammals, because of technology, which has eliminated the need for lactation and may well, in the future, eliminate the need for pregnancy. But what men lack is opportunity, and they lack it for the same reason that male langurs lack the chance to fondle newborn babes: They have been socially excluded by the behavior of the females. Women themselves, by culturally controlling the access of men to births, to infants, and to young children, have excluded them from the contact that apparently fosters "maternalistic" feelings seen in other mammals. By doing so, women have reinforced the myth that biology is destiny and that the female of our species is the only one who should care for the children.

Beginning with birth, women dominate the entire process of raising children and create their own old-girl network, which rivals that of males in terms of exclusivity, if not in terms of social power. All such power plays begin with the brain, but not in the way you may think.

Our incredible brains, so much larger than those of the other primates (who in their own right have large brains relative to most mammals) are of course responsible for the extraordinarily large size of the newborn human infant and the tight squeeze of its passage down the birth canal. However, that brain would have been an evolutionary impossibility if human society had not been able to assist women in labor. No other species has to have assistance, because no other species faces the problem of giving birth to a baby that is just too big.

Midwifery and obstetrics are skilled crafts in our cultures, but their beginnings may have been as crude as the technique of the male marmoset who manually aids his mate during a particularly difficult labor.

Childbirth is a social event — and fascinating to boot. As we mentioned, the fascination is not lost on other species. Sociable monkeys of both sexes are hypnotized by birth and its product; beaver fathers, too, join in the primal cleaning of the membrane-wrapped infant. But oddly, not all human adults have been allowed to participate in the essential social event of birth, and those excluded are almost always male. Over 70 percent of human societies ban men from observing or assisting their wives in birth. Birth is treated in the same way as menstruation among the Arapesh of New Guinea. The parturient woman is banished to the women's huts outside of the village, and her husband is not allowed to witness the birth; if he does witness it, his crops and hunting ability will be damaged. However, he moves in with the mother and infant soon after birth.

In the Nyansongo tribe of Kenya, the parturient female retires to her mother-in-law's house and is surrounded by a crowd of female relatives. Her husband and other men are barred from entering the house; however, the husband is put on call to dig up a special root if the delivery is difficult and lengthy. In Western society, until just recently, hospitals banned men from accompanying their wives into the delivery room. Their function was solely to pace the waiting room floor.

Given the diversity of human beliefs and procedures involving birth, it is astounding that so many societies agree on one thing: Men have no place at the birth. Women traditionally surround themselves with other women while they go through labor, but the exclusion of their husbands, who have such a big stake in the outcome, is remarkable. It seems as though women, quite obviously aware of their unique power over one aspect of human life — childbirth — conspire to extend that power beyond parturition. By leaving men in the passive role of anxiously awaiting news of the delivery — pacing in whatever special waiting room their society provides, mystified all the more by having no direct knowledge of what is going on — the female's power is complete.

Furthermore, many cultures demand that men stay away from

their wives and new babies for a period of days or even weeks following the birth. Proceeding from the events associated with childbirth, most women, by choice, become the caretakers. Whatever neural switches had to be turned on in order to care for the wrinkly, squashed, decidedly unattractive bundle of wailing that is a day-old infant were turned on in the brain of the mother but not the father. The combination of parturition hormones and constant contact with her new baby couldn't fail to evoke a maternal response from Mother, no matter how exhausted, depressed, and mangled she felt from the ordeal of labor; just like a mother rat, a human mother is urged by the environment to care for her baby. And just like a father rat, the human father is denied access to the environmental cues that would turn on his "maternal" behavior.

The father, not given the opportunity to be a caretaker, becomes less adept, while the mother becomes experienced at caretaking, improves her skills daily, and gradually becomes so proficient at caretaking that when the father is called upon for some small duty, his ineptness is glaring, leading some to comment that men just don't know how to care for children.

Because we would like to believe in the sanctity of motherhood as passionately as we believe in the sanctity of apple pie, we, as humans, may be loath to admit that maternalism, the tendency toward caretaking behavior, is not biologically programmed in the genes of the female. But as we have seen, the transient hormones of birth prime a female to begin to nurse. Soon afterward, however, these hormones begin to dissipate and the infant becomes the critical determinant of the mother's attention. Biologically primed, the mother-infant bond is actually a psychological one, emanating from the familiarity established between caretaker and offspring. Fathers and other adults are predisposed to caretaking as well. We could just as easily form father-infant bonds. Moreover, if the physiological changes associated with the only two true female monopolies, gestation and lactation, were necessary to maternalism, we would never be able to adopt an infant or give it the proper tender, loving care. Infant adoption, in fact, has been observed among other animals besides humans.

The creation of a caretaker is an infant strategy. By stimulating the caretaking tendencies of adults and by creating a bond, evolution assures the species that the infant will survive.

THE MYTH OF MATERNALISM

Under normal circumstances, the mammalian infant knows its mother best, for she must be with it at its birth and remains with it during its nursing phase. Clearly, if bonding is a species strategy, then a bond will form between mother and infant, giving rise to a belief in the exclusive rights of the mother as the source of caretaking behavior.

From the evidence concerning mammals and other vertebrates, we feel comfortable in concluding that care of offspring is not originated only by the caretakers. Offspring create caretakers too, through various behaviors, such as delicate mewing and tender nuzzling. Offspring stimulate nurturant, protective responses in the caretaker, beneficent to their survival—and ultimately to species survival.

9

NEW ROLES, NEW CHOICES

No trouble in their faces
Not one anxious voice
None of the crazy you get
From too much choice
The thumb and the satchel
Or the rented Rolls-Royce.
    —Joni Mitchell, "Barangrill"

THE FEMALES OF each of the millions of species of animals but one have no choice: They are only what biological evolution has made them, and they do what evolution decrees they do to be successful members of the species. For those females, that means making eggs, putting the eggs in the right place at the right time, and getting them fertilized. As we have seen, each species has its own way of accomplishing these feats, and each species takes care of its developing young in its own way, too. The ways in which a female rabbit, for example, with her eggs-on-demand system, accomplishes her reproductive mandate are quite distinct from those employed by a female fur seal, who experiences one brief day of fertility. Similarly, the parenting patterns etched into the brain of a monogamous female Canada goose are totally unlike those of a polyandrous female jacana.

    These females—the females of every species but one—lead reproductive lives marked by consistency. Evolution has shaped their appearance, their behavior, their physiology, and their mating and parenting strategies to complement one another. The female jacana, gaudier and more aggressive than her spouses, leaves to them the task of building the nest and incubating her eggs. Every facet of her strategy fits a life of polyandry. The female fur seal in heat, usually indifferent to the aggressive battles of the huge males, turns into a sex-hungry temptress who must have sperm for her briefly ready eggs; every aspect of her behavior suits her life of single parenthood.

The strategies of all females, considered together as a whole, trace the outline of what is "permissible" for females, of the leeway evolution allows her to accomplish her feminine goals. Within the large area of behavior that describes all females, however, any one species occupies only a small space. The fur seal can't extend her appetite for mates to longer than a one-day stand any more than the jacana can decide to settle down with a single mate until death do them part. They just don't have the choice.

Unique among all the millions of kinds of females in the world is the one creature who does have freedom of choice. It's no surprise that this privileged animal is the female of our own species. She is blessed with a biology that makes her different from other females in her capability to switch from one mating strategy to another. The biology that sets her apart — her lost estrus, her continuously low-level receptivity, her difficulties in childbirth, her menopausal cessation of reproduction — makes only the barest framework for the superstructure of human culture, which truly distinguishes her from other females.

The environment that we ourselves create — the rituals of our religions, the requirements of our laws, the technology that allows us to prevent conception and bottle-feed our babies — rules our reproductive lives with a firmer hand than does our own biology. It is the remarkable human invention of language that tells us, through the medium of information passed on from one generation to the next, what our reproductive strategies are to be. Words — not our hormones, and not some neurological gestalt etched into our brains — teach us how to find a mate, who is appropriate as a mate and who isn't, when to have babies, how to give birth to them, and how to take care of them.

Our species has taken full advantage of its flexibility by having no fixed, physiologically determined reproductive strategy. Many things other species do we also can do — if not better, at least as well. For example, women can mate monogamously for life, switch mates after a while, be harem members, or even rule a harem of males. And, indeed, they do, although not everywhere at once. Some societies

NEW ROLES, NEW CHOICES

permit polygamy, of either the polygynous or the polyandrous kind. Others, such as our own, limit legal matings to monogamous marriages, with an escape clause that permits divorce and remarriage. Still others frown upon any relationships but the lifelong monogamous kind. Similarly, expectations about child-rearing vary from one social setting to another. In some, both sexes are expected to take an active role in the day-to-day care of youngsters; in others, raising children is considered exclusively women's work. In some, the mother alone will nurse and otherwise tend the child; in others, breast-feeding is extended to relatives' children, and babies are handed around more freely. In yet other places, unrelated women may be hired to tend to, and even breast-feed, the kiddies.

But despite the freedom that language bestowed upon our species as a whole, individual women once had no more choice than a bird. Primitive societies preempted the right of making a choice of reproductive strategy (and still do, in some places), decreeing for women, just as evolution decreed for other females, what they were to do.

In other places and at other times, this system worked pretty well. Life didn't change much from one generation to the next, and the traditional rules, whatever they were, handed down from time immemorial, didn't go out of style. Women did women's work, and men did men's work, and both kinds of labor were essential to the society year in and year out.

But times have changed, and this system doesn't work so well in technological cultures. Technology should give to women choices undreamed of by most females. Women should be able to choose, with incredible certainty, if they will have children and when they will have them. They should be able to choose whether or not they will breast-feed their infants. Technology gives women control over their own reproductive strategy—something no other species has. But at the same time, technology has run headlong into the traditional controls exerted by society.

Freedom of choice is a mixed blessing. It should give women the chance to try new feminine roles, but it also has a schizophrenic, divisive effect. Like the person in Joni Mitchell's song, women are faced with a bewildering embarrassment of riches but with no guidelines as to how an individual can make a sane choice among the options. Should she stick with the traditional role, try to make it in the domain of men, or attempt to combine the two? Should she let males into the traditional female power structure of home and family as she tries to enter the male-dominated power structure, or should she be a Superwoman, running both home and office? Does she even want a male around to help share with raising the kids, or would she rather be a single parent?

Human society hasn't yet sorted out any of these, and many other, choices. Women are experimenting with new ways of accomplishing old goals, and the neatness, simplicity, and consistency that distinguishes the reproductive strategies of other females does not characterize our own. Indeed, the traditional woman's role in Western societies is rife with inconsistencies. For example, a woman traditionally dresses in brighter colors, dons more eye-catching jewelry and makeup, fixes her hair more elaborately, and wears more perfumes than a man, who favors drabber, more neutral accoutrements. This woman is unknowingly following in the footsteps of polyandrous females, who must compete among themselves for the attentions of the males. Culture has forced her to play a mating game in which she is "chosen" by the opposite sex, where the drab male is the one who gets to "pop the question." And yet the goal of this courtship ritual is not a harem of males, but rather, a long-term monogamous relationship. Where, then, is the reciprocal ritual and the similar appearance of the sexes of monogamous animals?

Perhaps as women experiment with new roles, perhaps as the limits to the possible are tested by both sexes, perhaps then freedom of choice in our reproductive strategies will not make people so troubled and anxious. The intent of this book has been to show that females fulfill their biological destiny in any number of ways and that humans should not feel that different sets of reproductive strategies are wrong or unnatural.

Oddly enough, that lesson is probably the only one we can learn from other animals. Theirs is a biological destiny; ours a hybrid of biology and culture, with the emphasis on culture. We will never be able to find among other animals the key to what we are now or might someday become; the best the animals can do is provide clues as to what we once were. Women will have to pioneer new female territory before they can find their own optimal strategies. That task will not be easy or straightforward, but it is the one that is our destiny.

CHAPTER NOTES

Chapter 1

Daley, Martin and Wilson, Margo. *Sex, Evolution and Behavior*. North Scituate, MA: Duxbury Press, 1978. Quotation in text from page 48.

Halliday, Tim. *Sexual Strategy*. Chicago: University of Chicago Press, 1982. Quotation in text from page 7.

Helvey, Mark. "First Observations of Courtship Behavior in Rockfish Genus *Sebastes*." *Copeia* 1982 (No. 4).

Wilson, E. O. *Sociobiology*. Cambridge, MA: Belknap Press of Harvard University Press, 1975. Quotation in text from page 318.

Chapter 2

LeGuin, Ursula K. *The Left Hand of Darkness*. New York: Ace Books, 1969. Quotation in text from pages 90–91.

Meisel, Robert L. and Ward, Ingeborg. "Fetal Female Rats are Masculinized by Male Litter Mates Caudally in the Uterus." *Science* 213 (1981): 239–242.

Money, John and Ehrhardt, Anke. *Man and Woman, Boy and Girl*. Baltimore: Johns Hopkins University Press, 1972. Quotation in text from page 99.

Vom Saal, Frederick and Bronson, F. H. "Sexual Characteristics of Adult Female Mice Are Correlated with Their Blood Testosterone Levels During Prenatal Development." *Science* 208 (1980) 597–599.

Chapter 3

Mead, Margaret. *Male and Female*. New York: William Morrow and Company, 1949. Quotations in text from pages 169–170 and 176.

Mead, Margaret. *Sex and Temperament in Three Primitive Societies*. New York: William Morrow and Company, 1935.

Mitchell, Gary D. *Behavioral Sex Differences in Non-Human Primates*. New York: Van Nostrand Reinhold Company, 1979. Quotation in text from page 81.

Tyler, Stephanie J. "The Behaviour and Social Organization of the New Forest Ponies." *Animal Behaviour Monographs* 5 (1972): part 2.

Vandenbergh, John and Drickamer, Lee. "Reproductive Coordination Among Free-ranging Rhesus Monkeys." *Physiology and Behavior* 13 (1974): 373–376.

Chapter 4

Darling, Frank Fraser. *A Herd of Red Deer.* London: Oxford University Press, 1956. Quotation in text from pages 177–178.

Hinde, R. A. "Interaction of Internal and External Factors in Integration of Canary Reproduction." In F. A. Beach (ed.), *Sex and Behavior*. New York: Wiley, 1965.

Lehrman, Daniel S. "The Reproductive Behavior of Ring Doves." *Scientific American* November (1964): 82–88.

Chapter 5

Alexander, Richard D. and Noonan, Katherine M. "Concealment of Ovulation, Parental Care, and Human Social Evolution." In N. Chagnon and W. Irons (eds.), *Evolutionary Biology and Human Social Behavior*. North Scituate, MA: Duxbury Press, 1979. Quotations in text from pages 442 and 453.

Beach, Frank. "Human Sexuality and Evolution." In W. Montagna and W. A. Sadler (eds.), *Reproductive Behavior*. Plenum, NY: Plenum Press, 1974.

Benshoof, Lee and Thornhill, Randy. "The Evolution of Monogamy and Concealed Ovulation in Humans." *Journal of Social and Biological Structures* 2 (1979): 95–106.

Burley, Nancy. "The Evolution of Concealed Ovulation." *The American Naturalist* 114 (1979): 835–858.

Galdikas-Brindamour, Biruté. "Orangutans, Indonesia's People of the Forest." *National Geographic* 148 (1975): 444–473. Quotation in text from page 470.

Hrdy, Sarah Blaffer. *The Woman That Never Evolved*. Cambridge, MA: Harvard University Press, 1981.

Morris, Desmond. *The Naked Ape*. New York: Dell, 1969.

Nadler, Ronald D. "Sexual Cyclicity in Captive Lowland Gorillas." *Science* 189 (1975): 813–814. Quotation in text from page 813.

Stone, Lawrence. "Starting with Eve." Review of *A History of Women's Bodies* by E. Shorter in *New York Times Book Review*, January 2, 1983, pages 6–11. Quotation in text from page 11.

Chapter 6

Buechner, Helmut and Schloeth, Robert. "Ceremonial Mating Behavior in Uganda Kob." *Zeitschrift für Tierpsychologie* 22 (1965): 209–225.

Spieth, Herman. "Evolutionary Implications of Sexual Behavior in Drosophilia." *Evolutionary Biology* 2 (1968): 157-193.

Thornhill, Randy. "Sexual Selection in the Black-tipped Hangingfly." *Scientific American* 242 (1980): 162-172.

Vandenbergh, John and Drickamer, Lee. "Reproductive Coordination Among Free-ranging Rhesus Monkeys." *Physiology and Behavior* 13 (1974): 373-376.

Welty, Joel Carl. *The Life of Birds*, 2nd ed. Philadelphia: W. B. Saunders Company, 1975. Quotation in text from page 262.

Wiley, R. Haven. "Territoriality and Non-random Mating in Sage Grouse *Centrocercus Urophasianus*." *Animal Behaviour Monographs* 6 (1973): part 2.

Wilsson, Lars. *My Beaver Colony*. Garden City, NY: Doubleday, 1968.

Chapter 8

Hrdy, Sarah Blaffer. *The Woman That Never Evolved*. Cambridge, MA: Harvard University Press, 1981.

Maccoby, Eleanor and Jacklin, Carol Nagy. *The Psychology of Sex Differences* Volume 1. Stanford: Stanford University Press, 1974. Quotations in text from pages 215 and 354.

Mitchell, Gary D. *Behavioral Sex Differences in Non-Human Primates*. New York: Van Nostrand Reinhold Company, 1979. Quotations in text from pages 204 and 210.

Mitchell, Gary; Redican, William; and Gomber, Jody. "Males Can Raise Babies." *Psychology Today* April 1974.

Oakley, Ann. *Sex, Gender and Society*. New York: Harper and Row, 1972.

Rosenblatt, J. et al. "Progress in the Study of Maternal Behavior in the Rat: Hormonal, Nonhormonal, Sensory and Developmental Aspects." *Advances in the Study of Behavior* 10:226.

Shaw, E. and Innes, K. "The 'Infantilization' of the Cyclid Fish." *Developmental Psychobiology* 13 (1980), No. 2.

Skutch, Alexander. *Parent Birds and Their Young*. Austin: University of Texas Press, 1976. Quotation in text from page 363.

Welty, Joel Carl. *The Life of Birds*, 2nd ed. Philadelphia: W. B. Saunders Company, 1975. Quotation in text from page 248.

Woolfenden, Glen. "Florida Scrub Jay Helpers at the Nest." *Auk* 92 (1975): 1-15.

INDEX

Abalones, 92
Acorn woodpeckers, 135
Adders, 106
Adrenogenital syndrome, 37
Aggression, 104
Alexander, Richard D., 81–82, 83, 86
Alligators, 40, 122–23
Alor (people), 142
Androgyny in literature, 45
Anemia, 127
Angler fish, 16–17
Anis (bird), 135
Anolis lizards, 114
Anopheles mosquitoes, 20–21
Anorexia nervosa, 48
Aphids, 22–23
Arapesh (New Guinea people), 55, 147
Arctic terns, 135
Arrenta (people), 142
Artificial insemination, 71
Asexual reproduction, 19–24
Atlantic silversides, 39

Baboons
    matriarchy among, 3
    offspring of, 144–45
Bacteria, 24–25
Baleen whales, 16
Barr body, 30
Bats, 16
Beach, Frank, 75–76, 86
Beagles, 75–76
Beavers, 118, 143
Behavior
    biological determinism and, 2
    choice regarding, 151–54
    courtship, *see* Courtship behavior
    hormone levels and, 4
    proceptive, 76
    sexual, *see* Sexual behavior
    stereotypical expectations of, 2
Benshoof, Lee, 82–83, 84
Birds

courtship behavior in, 97–99
egg incubation by, 124
egg laying genetically programmed in, 61
incubation physiology of, 131
male parenting among, 135–36
mating behavior of, 61–63, 115–19
nest-building behavior among, 63, 115–17
offspring care in, 134–35
pair bonding among, 114–17
polyandrous, 11
seasonal sexuality of, 49
songs of, 113
territory acquisition among, 15
weed gathering by, 115–16
See also specific birds
Birds of paradise, 114
Birth, 130–31
    cultural attitudes and, 147–48
    of mammals, 129–30
    of shiner perch, 10–11
    as social event, 147
Birth control, 71–72, 85
Black-tipped hanging fly, 111–12
Black widow spider, 2
Blood, embryo relationship to, 127
Blue acara fish, 123
Blue whales, 16
Bluebirds, 134
Bonellia (echiuroid worm), 17, 40–41
Bororo (people), 142
Bower birds, 102–3, 135
Brain
    birth and, 146
    body fat monitored by, 47
    gender and function of, 32–33
    hormones and, 33–34
Breast-feeding, 142
Bronson, F. H., 35
Brown, Helen Gurley, 88
Buechner, Helmut, 100
Burley, Nancy, 84–86
Bushtits, 134

Butterflies, 11, 110–11

Camels, 70
Canada geese, 116–17, 151
Canaries, 63
Capuchin monkeys, 144
Catfish, 124
Cats, 5, 6–7, 51, 70
Childbirth, see Birth
Child-rearing, 138–39, 142–49, 153
Chimpanzees, 11, 76–77, 80, 87, 145
Choice, reproductive, 151–54
Chromosomes, 28–32
Cichlids, 124–25, 136, 141–42
Clams, 15, 92
Cleaner wrasses, 114
Clown fish, 44
Coatis, 142–43
Color blindness, 29–30
Contraception, 71–72, 85
Copeia (journal), 13
Cottids, 136
Courtship behavior, 91–120
   bird weed-gathering in, 115–16
   female approach to male in, 102–7
   of insects, 109–12
   pair bonding in, 114–20
   preparation time in, 113–14
   role of lek in, 96–102
   of sessiles, 92–93
   sex bargaining in, 110–14
   sound in, 113, 115
Cows, 37–38, 137
Crepidula fornicata, 43
Crickets, 114
Cuckoos, 134, 135
Cupiennius salei, 106–7

Dakota Indians, 142
Daley, Martin, 13
Daphnia, 2, 21–23
Darling, F. Fraser, 66–67
Deafness, 29
Dik-dik, 128
Division of labor in child-rearing, 138–39
Dragonflies, 114
Drickamer, Lee, 103–4
Drosophilia, 11, 100–101

Duck-billed platypus, 128
Ducks, 135
Dysplasia, 29

Echinoderms, 121
Eggs, 13–14, 26–27
   bird, 62–63
   birds' incubation of, 124
   carrying of, 124–26
   fertilization of, 121–31
   fish, 64–65
   genetic materials in, 28
   guarding of, 122–24
   life expectancy of, 70
   male guarding of, 107–10
   of monotremes, 128
   nutrients in, 126
   of shellfish, 92–93
   size of, 28, 127–28
   of snakes, 106
Ehrhardt, Anke, 37
Elephants, 50, 77, 143
Elk, 130
Embryos, 121–31
   nurturing of, 126–27
   as parasites, 126–30
   sex differentiation in, 32–41
Endocrine system, embryos' effect on, 127
Environment, reproduction role of, 59–60
Estrogen
   composition of, 53
   menopausal decrease in, 73
Estrus, 75–89
   human concealment of, 78–89

Fat, 47–48
Fathers, exclusion in child-rearing of, 144–49
Female interconnectedness, 1
Femininity, evolutionary basis of, 4
Ferrets, 70
Fertility, techniques for improving, 72
Fertilization, 121–31
   sperm function in, 28
Fiddler crabs, 114
Field voles, 49, 50
Fin whales, 49

INDEX

Fish, 15–16
　egg carrying by, 124–25
　egg laying by, 137
　livebearing, 125
　male egg guarding among, 107–10
　male offspring guarding among, 136–37
　parenting in, 123–24
　sex determination in, 39–40, 43–45
　spawning in, 93–94
　See also specific fish
Follicle-stimulating hormone, 56, 73
Foxes, 68
Frogs, 113, 114, 122
Fruit flies, 11, 100–101
Fur seals, 104, 152

Galdikas-Brindamour, Biruté, 76
Gametes, see Eggs; Sperm
Garter snakes, 11
Gasterosteids, 136
Geese, 116–17, 118
Gender
　acculturation and, 45–46
　brain function and, 32–34
　chromosomal development of, 27–32
Genetics
　chromosomal function in, 27–32
　mutation in, 38
　variation in, 24–27
Gerbils, 51
Gibbons, 79
Gift giving, 112
Glaucoma, 29
Goats, 137
Gomber, Jody, 141
Gonadectomy, 71
Gonads, 33–35
Gorillas, 2, 52, 76, 87
Grebes, 15, 115–16
Grizzly bears, 2
Grosbeaks, 134
Groupers (fish), 44–45
Guinea pigs, 87
Guppies, 125

Habrobracon, 53–54
Halliday, Tim, 14
Hamsters, 16
Harem master, stereotype of, 105–6
Hares, 2–3
Hemophilia, 29
Herpes virus, 102
Herrings, 93
Hinde, Robert, 63
Honeybees, 42
Hormones
　behavior and, 4
　brain function and, 33–34
　communication mechanism of, 54
　disruptive influence of, 36–37
　embryo manipulation of, 127
　fat and, 47
　shiner perch sexuality and, 8
　influence on embryos by, 35–36
　placental regulation of, 129
　pubescent changes in, 51–58
　sensory mechanisms modified by, 78
Horses, 130
Howler monkeys, 144
Hrdy, Sarah Blaffer, 84, 142
Hurler's syndrome, 29
H-Y antigen, 32, 35, 38
Hyenas, 17
Hypothalamic-pituitary center, 52, 53, 56
Hypothalamus, 33, 34

Identity, maintenance of, 1
Incubation
　by birds, 124
　by monotremes, 128
Insects, courtship behavior in, 109–12

Jacana birds, 12, 135–36, 151
Jacklin, Carol, 138–39, 146
Jays, 134–35
Jesus birds, 15, 135–36
Jewel fish, 5, 123

Kangaroos, 114
Katydids, 111
Killdeer, 136

Kingfishers, 134
Kobs, Uganda, 99–100

Lactation, 133, 137, 138
Langurs, 145
LeGuin, Ursula K., 45
Lehrman, Daniel, 61–63
Lek, 96–102
Lemon tetras, 113
Leopard frogs, 70
Life-styles, biological and cultural factors determining, 3
Lions, 142
Lobsters, 64–65
Lowland gorillas, 52, 76
Luteinizing hormone, 56–57, 73

Maccoby, Eleanor, 138–39, 146
Magpie geese, 134
Malaria, 20–21
Male mandate, myth of, 5–7, 11–18, 104
Mallards, 135
Manus (Pacific people), 55
Marmosets, 131, 144
Maternalism
    female suitability in, 138–39
    male potential for, 146
    mother-infant bond, 137
    myth of, 133–52
Maternity, sharing of, 142–44
Matriarchy among baboons, 3
Mead, Margaret, 55, 142
Meisel, Robert, 36
Menopause, 72–73
Menstruation, 52–58
Mice
    estrus behavior in, 35
    puberty in, 50
Milk, 133, 137, 138
Minks, 70
Mitchell, Gary, 52, 141, 146
Mitchell, Joni, 151, 153
Moles, 70
Molluscs, 121
Monarch butterflies, 110–11
Money, John, 37
Monkeys, child-rearing among, 142
Monogamy, 153
    sociobiology's theory of, 81–84
Monotremes, 128

Moon, mating timed to phases of, 92
Moose, 48
Morris, Desmond, 79, 83
Mosquitoes, 11, 20–21
Mothering, see Maternalism; Maternity
Mouthbrooder fish, 64
Mule deer, 48
Murres, 134
Mussels, 92

Nadler, Ronald, 76
Nest building by birds, 63, 115–17
New Forest mares, 51–52
Newts, 64
Night blindness, 29
Night monkeys, 144
Noonan, Katherine M., 81–82, 83, 86
Norway rat, 5
Nursing, 137, 138, 142, 143
Nurturing of embryos, 126–27
Nuthatches, 134
Nutrition, puberty age relationship to, 48
Nyansongo (Kenyan peoples), 147

Oakley, Ann, 142
Octopuses, 123
Old Testament, 55
Orangutans, 76
Ovum, see Eggs
Oysters, 15, 27, 92

Pair bonding, long-term, 114–20
Palolo worms, 92
Paramecia, 19
Parasites, 20–21
    embryos as, 126–30
Parthenogenesis, 22–23
Paternity, 135–36
Pelicans, 15
Penguins, 134
Phalaropes, 15, 136
Pheromones, 78
Pied kingfishers, 134
Pigeons, 134
Pigs, 75, 143
Pinnipeds, 104–6
Pituitary gland, 33

INDEX

Placenta, 128–30
Plant lice, 22–23
Plasmodium, 20–21
Platyfish, 125
Polar bears, 128
Polyandry, 84, 152, 153
Polygamy, 153
  sociobiology's theory of, 81–84
Pregnancy, 59–73
  anemia during, 127
  contraception and, 71–72
  cultural attitudes toward, 85–86
  seasonal relationship to, 49
  of shiner perch, 10–11
  teenage, 48
Proceptive behavior, 76
Progesterone, composition of, 53
Promiscuity, 11, 109
Protein, butterfly consumption of, 110–11
Protozoans, 26
Puberty, 47–58
  body fat and, 47–48
  hormonal changes during, 51–58
  menstruation during, 52–58
  nutrition and, 48
  seasonal relationship to, 49
  social interaction as factor in, 49–50
Pupfish, 108–9
Pygmy nuthatch, 135

Queen bees, 42

Rabbits, 16
  estrous cycle of, 69–70
Rats, 5, 15, 130
  estrous cycle of, 68–69
  nursing among, 143
  offspring of, 139–41
  testosterone levels of, 34, 36
Rays, 126
Red deer, 48, 137
  reproductive behavior of, 66–69
Redican, William, 141
Redwing blackbirds, 114
Religion, menstruation and, 55
Reproduction, 59–73
  asexual, 19–24
  choice in, 151–54
  chromosomal function in, 27–32
  egg production time and, 70–71
  environment and, 59–60
  genetic variation provided by, 24–27
  in red deer, 66–69
  in shiner perch, 10–11
Rhea (bird), 136
Rhesus monkeys, 52, 58, 77, 80, 87, 103–4, 130, 141
Ring doves, 61–63
Roman Catholicism, 89
Rosenblatt, Jay, 140–41

Sagegrouse, 97–98, 135
Salamanders, 2, 11, 64, 125
Samoans, breast-feeding by, 142
Schloeth, Robert, 100
Science, male bias of, 3, 5–7, 11–18
Sculpins, 112, 123
Sea lions, 105
Sea turtles, 40
Sea urchins, 15, 92
Seahorses, 125
Seals, 104–6
Sergeant-major fish, 123
Sessile females, mating in, 92
Sex role reversal, 11–18, 136
Sexual behavior
  myth of male mandate in, 5–7, 11–18, 104
  promiscuity in, 11, 109
  of shiner perch, 8–11
Sexual stimulants, smell as, 51–52, 75, 77
Sexual union as genetic engineer, 25
Sharks, 126
Shaw, Evelyn, 141–42
Sheep, 49, 77, 130, 137
Shellfish, mating behavior in, 92–93
Shiner perch, 3–4, 118
  habitat of, 9–10
  pregnancy and birth in, 10–11
  sexual behavior of, 8–11
  sperm storage by females, 10
Singles bars, lek analogy of, 101–2
Size, mate selection on basis of, 112
Skutch, Alexander, 134
Slipper shell, 43
Smell as sexual stimulant, 51–52, 75

Snails, 15, 43
Snakes, 11, 106, 107
Society
  child-rearing attitudes of, 138–39, 142–49
  new choices for, 151–54
Sociobiology
  estrus loss explanation of, 80–86
  sexist nature of, 12–13
Songs of birds, 112, 115
Sound, courtship behavior and, 113, 115
Sparrows, 5, 113
Sperm, 13–14, 26–27
  breeding season production of, 91–92
  of butterflies, 110–11
  conservative anatomy of, 27–28
  female inducement to production of, 103
  fertilization function of, 28
  of honeybees, 42
  of shiner perch, 8, 10
Spiders, 2, 16, 106–7
Spieth, Herman, 101
Spiny anteater, 128
Squid, 94–95
Squirrel monkeys, 144
Squirrels, 11
Starfish, 15
Starworm, 40–41
Stereotypes
  behavioral, 2
  in scientific research, 3, 11–18, 136
Sticklebacks, 123
Stone, Lawrence, 86
Superstition, menstruation and, 54–56
Surf perches, 126
Swallows, 134
Swans, 116–17, 118
Swifts, 134
Swordtail fish, 11, 125

Tamarin monkeys, 144
Tanagers, 134
Technology, 1, 138, 153
Terns, 134
Testicular feminization, 38

Testosterone
  fetal excess of, 37
  rat behavior and, 34, 36
Thornhill, Randy, 82–83, 84, 111–12
Three-spined stickleback, 107–8
Tinamous (bird), 136
Tit monkeys, 144
Toads, 113, 125
Tree frogs, 122
Troop monkeys, 131
Turner's syndrome, 38–39
Tyler, Stephanie, 51

Uganda kob, 99–100
Urine
  courtship function of, 100
  as sexual stimulant, 51–52, 77

Vandenbergh, John, 50, 103–4
Venereal disease, 102
Vitamin C, 87
Vom Saal, Frederick, 35

Walruses, 104–6
Wapiti elk, 48
Ward, Ingeborg, 36
Wasps, 53–54
Water fleas, 2, 21–23
Waterbugs, 109–10
Weaver birds, 114
Weed gathering by birds, 115–16
Welty, J. C., 103, 136
Whales, *see specific whales*
Wild-dogs, 143
Wiley, R. Haven, 98
Wilson, E. O., 12–13
Wilson, Margo, 13
Wilsson, Lars, 118
Wood ducks, 135
Woodpeckers, 134, 135
Woolfenden, Glen, 134–35
Wrens, 112, 134

X chromosomes, 29–30

Y chromosomes, 29–32

Zebra butterflies, 110–11
Zygote, 25, 26, 28, 31

Outstanding Paperback Books from the Touchstone Library:

☐ **The New Our Bodies, Ourselves**
by The Boston Women's Health Collective
The completely revised, updated, and expanded edition of the most important book to come out of the women's movement. This complete, reliable, courageous sourcebook on women's health care for all generations now has new chapters on Body Image; Alcohol, Drugs, and Smoking; Psychotherapy; Violence Against Women; New Reproductive Technologies; Women Growing Older; and more.
46088-9 $14.95

☐ **Women in Science**
by Vivian Gornick
How has the role of women in science affected the field as a whole? In this fascinating book, Vivian Gornick, noted author of *Essays in Feminism,* explores the emotional, intellectual, and professional experiences of women struggling for recognition in the world of scientific research.
41739-8 $6.95

☐ **A New View of a Woman's Body**
by The Federation of Feminist Women's Health Care Centers
This fully illustrated guide contains vital information on self-examination, reproductive anatomy, birth control, menstrual extraction, abortion care, menopause, and other health issues; it demystifies women's bodies, redefines the physiological aspects of women's sexuality, and promotes a new understanding of women's health care.
41215-9 $10.95

☐ **Originals: American Women Artists**
by Eleanor Munro
Finally, American women artists of the 20th century are celebrated in a fascinating, fully illustrated work of art history and criticism. Munro's interviews and research bring us the lives and work of the late O'Keeffe, Krasner, Frankenthaler, Nevelson, and many other remarkable artists. "It is passion that makes this book shine."
—*The New York Times Book Review*
42812-8 $15.95

☐ **Deciphering the Senses: The Expanding World of Human Perception**
by Robert Rivlin and Karen Gravelle
We may actually possess as many as *seventeen* senses. Drawing on the most recent research, the authors discuss the "classic five," other amazing sensory organs, pain, perception, and illusion in a work that *Chicago Booklist* called "good reading on a fascinating and inherently interesting topic."
46124-9 $7.95

☐ **The Universe Within: A New Science Explores the Human Mind**
by Morton Hunt
Morton Hunt invites us to look into the dazzling mystery and radical uniqueness of human intellectual ability. "Uncommonly readable... exhilarating..."—*Ashley Montagu*
25259-3 $10.95

MAIL THIS COUPON TODAY—NO-RISK 14-DAY FREE TRIAL

Simon & Schuster, Inc.
Simon & Schuster Building, 1230 Avenue of the Americas,
New York, N.Y. 10020. Mail Order Dept. FH1

Please send me copies of the above titles. (Indicate quantities in boxes.)
(If not completely satisfied, you may return for full refund within 14 days.)

☐ Save! Enclose full amount per copy with this coupon. Publisher pays postage and handling; or charge my credit card.
☐ MasterCard   ☐ Visa

My credit card number is _____ Card expires _____
Signature _____
Name _____
(Please Print)
Address _____
City _____ State _____ Zip Code _____
or available at your local bookstore   Prices subject to change without notice